After They're Gone

Extinctions: Past, present and future

After They're Gone

Extinctions: Past, present and future

PETER MARREN

HODDERstudio

First published in Great Britain in 2022 by Hodder Studio
An Hachette UK company

1 3 5 7 9 10 8 6 4 2

A CIP catalogue record for this title is available from the British Library

Hardback ISBN 9781529393408
eBook ISBN 9781529393415
Audio ISBN 9781529393422

Typeset by Palimpsest Book Production Ltd, Falkirk, Stirlingshire

Printed and bound in Great Britain by Clays Ltd, Elcograf S.p.A.

Hodder & Stoughton policy is to use papers that are natural, renewable
and recyclable products and made from wood grown in sustainable forests.
The logging and manufacturing processes are expected to conform to the
environmental regulations of the country of origin.

Hodder & Stoughton Ltd
Carmelite House
50 Victoria Embankment
London EC4Y 0DZ

www.hodder-studio.com

'Extinction is the rule. Survival is the exception.'
Carl Sagan, *The Varieties of Scientific Experience*, 2006.

'We defy augury. There is special providence in the fall
of a sparrow. If it be now, 'tis not to come. If it be not
to come, it will be now. If it be not now, yet it will
come. The readiness is all.'
William Shakespeare, *Hamlet*, Act 5, scene 2.

To Tim Birkhead, Michael McCarthy and Jeremy Mynott

In gratitude, friendship and esteem.

Contents

Introduction:
Different ways of dying out

Moriendum enim certe est; et id incertum an eo ipso die.
Death is inevitable; you just don't know exactly when.
Cicero, De Senectute, 74.

Extinction gets us all in the end. Ninety-nine per cent of every life form that has ever existed is extinct. Every species on earth today, including *Homo sapiens*, will one day be extinct. In some cases it may be quite soon. For a few it will be today. Extinction is final – the most final thing of all – yet it is not necessarily tragic. It's more of a fact of life, like breathing, like a heartbeat. Hearts eventually stop. Animals and plants die, and when death exceeds new life, then extinction is on the way. And, biologically speaking, death is just as important as life. Endings are also beginnings. We move on. Life moves on.

The tragedy lies more in what we have done to extinction. It was a natural process that we have made unnatural. We have pushed the evolutionary accelerator and maximised the revs. Extinction is out of control. It is turning into a mass extinction of life, only the sixth such event in the living planet's long history. This man-made extinction will haunt us for as long as we live, until we become extinct too.

Yet the extinction of species is easy enough to ignore. Only occasionally does an example hit the headlines, such as the apparent extinction of the baiji or Yangtze river dolphin in 2007 – newsworthy as the first species of the world's megafauna to die out for decades. Or the loss of the beautiful Spix's macaw, a startlingly beautiful, bright blue parrot, declared extinct in the wild in 2019, although it still survives in captivity. But how many of us know about the kouprey, or Cambodian forest ox, last seen about 1970, or Schomburgk's deer, all shot to death by 1932 for their magnificent antlers, or the Christmas Island forest skink, gobbled up by an introduced snake and all gone by 2021? Some of the names of dying species seem eerily prescient: the gloomy tube-nosed bat; the Sulu bleeding-heart; the Bermuda flicker; the lost shark.

Yet most declared extinctions are not picked up by the press. They are barely noticed by anyone except specialists. And, very likely, most of the extinctions that are taking place will be of animals and plants that haven't been discovered yet: insects, worms, fungi, micro-organisms. We can only guess at the number of species vanishing from the earth every year, entirely unnoticed, gone before the scientists could get round to naming them. If you are a vertebrate biologist, you will probably know the sad stories of the Japanese sea lion or the toolache wallaby. If you twitch birds, you will probably realise that you are never going to add the paradise parrot or the Labrador duck to your life list. If you love insects, you have probably heard of the Xerces blue, the world's most famous extinct butterfly (but only because it once lived in the Sunset district of San Francisco). But most species pass away unknown and unmourned. The planet is said by some to be dying but who can name the dead? Where are all the bodies? Without named examples we are only raging about a statistic, a probability.

In this book, I hope to show you quite a lot of bodies. And yet more animals and plants that are on the way out and likely to join them before too long. Join them, that is, unless we change our ways, cease to exploit vulnerable animals and plants, protect their habitats and leave an adequate space for them to exist. How likely that will be, I will leave you to decide. We must live in hope, but it seems to me that most environmental policies are predicated on hope alone. It is as if hope alone will do the work, guide us to a better future. The trouble with that is that we are human. Sure, we want to save the planet (though we might argue about what we mean by 'save') but we also want to enjoy ourselves and earn enough to have two cars, a nice house, holidays abroad and a family.

How do we witness extinction? How do we know when a species has vanished? Lists of newly extinct species are compiled annually by the International Union for Conservation of Nature (IUCN), and the data as a whole are reassessed every ten years. These annual lists are 'declared' extinctions. They are the species that a scientific consensus has concluded are no longer around. Most of them were in fact last seen alive many years ago. It takes time for certainty to descend. Lost species have to be looked for. For tiny things like insects or snails, likely extinction condenses into a certainty only when the species happens to inhabit a confined space that can be thoroughly searched, such as a small island or an isolated habitat. That is what happened to the Xerces blue butterfly after its entire living space, its home on the wild Pacific dunes, was transformed into Sunset District and the Golden Gate Park.

The actual moment of an extinction of course takes place in an eye-blink as the last representative of its kind on earth passes away. It will be witnessed only when an animal dies in captivity, such as the very last passenger pigeon, called Martha, who died on 1 September 1914, between midday and one o'clock, at Cincinnati Zoo in Ohio. Otherwise our sense of it has to be retrospective. But while the actual moment of extinction is an instant, the process towards oblivion often takes a long time, a slow, gradual descent that may be barely noticed within a human lifetime (or a human memory). Some species may decline and then recover. Others oscillate, with good years – or decades or centuries – followed by bad ones. Such a species may be in long-term decline without anyone realising it.

Anyone who loves nature will know of once-familiar animals or birds or wild flowers that are disappearing from our neighbourhoods, species that seem to get rarer year on year, until one day they are no longer our neighbours, and so no longer part of our lives. I live in a large village in the Kennet valley, in Wiltshire, right in the middle of southern England. Twenty years ago a nightingale sang from the overgrown hedge opposite my bedroom and I am probably one of the few in England today who has been kept awake by John Keats' immortal bird, chortling in 'full-throated ease' all night long. But the nightingales have gone, and so have the snipe that used to drum over the wet meadows by the river, as have the lapwings whose wheezy calls announced the coming of summer. They have gone not only locally but in great swathes across England. Efforts are being made to ensure they will not become extinct nationally. Or, at the very least, to slow down the inevitable.

England is said to be one of the most nature-depleted countries in the world. It has lost most of its big wild animals. A thousand years ago in my village, which was quite a big place even then, beavers used to build their dams on my river. Perhaps a thousand years before that the villagers might have heard a wolf howl from a distant down or a rumour of bears in the forests that extended for mile after mile. Even a hundred years ago, there would have been far more wild flowers, particularly at the edge of the cornfields, colourful blooms whose names attest to their former familiarity: shepherd's needle, corn cockle, poor man's weather-glass.

A tiny plant called a spike-rush grows in the meadow opposite my house. It is a minimalist plant, barely more than a stalk with a rush-like flower on top, and superficially resembling some of the first land plants on earth. I thought that there was something odd about this particular spike-rush, and an expert who examined it for me, probably under a powerful microscope, told me it was probably a genetic mix, not a hybrid exactly, but a plant that had absorbed some of the characters of another, related species, a now locally extinct one called the needle spike-rush. And, if so, that tiny lost species represented only in the genome of another plant, is an echo of what was once here but has long since gone: a genetic whisper as it were, a ghost-plant. On a scale from 1 to 10, with our lost beaver at 10, I'd give the needle spike-rush about 0.1. But it makes a point. One's direct experience of extinction is likely to be undramatic. Even mass extinction happens slowly, so slowly that hardly anyone notices unless a species happens to be studied and monitored, as birds and butterflies are in Britain.

Extinction represents the sharp end of a slow bleeding of biodiversity. It is incremental, arriving eventually at the solitary pole of permanence in an otherwise fluid and potentially reversible process. We can never know whether a species will become locally extinct until it has gone. For instance, how many of Britain's (and Northern Europe's) ash trees will survive? Most are now dying from a fungal disease as part of the price we pay for the lack of import controls on timber or nurseling trees. Will some survive? Will the tree eventually recover? We don't know. We might notice that grasshoppers no longer spring from our feet as we walk through the grass in midsummer. Are they on the slide? People talk about the lack of flying insects in the car headlights, or on the ceiling when they leave the window open at night. When our river was poisoned after someone tipped insecticide down the drain, the mayflies suddenly vanished and, although some have since returned, others haven't. Yet one's perception of nature is barely dented by these disasters. The birdsong in spring is still spectacular – the sweetly warbling blackcaps, the fluting blackbirds, the cuckoo's wandering voice. Each spring you appreciate anew nature's extraordinary resilience, its ability to find refreshment in a half-empty vessel. Nature, I think, is an optimist.

But is *Homo sapiens* turning into a pessimist? Each year, we are seeing the terrifying results of a rapidly changing climate. Where I live, we receive more rain than before, especially in winter when what used to be a fairly firm path across the meadows is now a porridge of slippery mud. Winter winds are stronger. Trees that had managed to stay upright for hundreds of years are toppling. Instead of the traditional seasons we seem to get alternate seams of flood and drought. And in place of winter snow there are endless, endless grey skies. Nature is in flux, and in turbulent times like these there will be winners and losers. I see some of the winners from my window, the little egrets alighting like snow-white shawls and red kites, the progeny of releases in the 1990s, floating overhead with a sharp eye for road-kill. But on the other hand I cannot now expect to hear the nightingale, nor the soft crooning of the turtle dove. They have left us. They are only memories.

Like, perhaps, most naturalists I would prefer things to stay the same. Politicians tell us that change is an opportunity. Yes, but unfortunately all too often it is an opportunity for extinction.

Wildlife is losing. It is not their fault. All species on earth are well-adapted to their respective environments and have survived until now. It is we that have taken away what supported them, what they need most of all: their homes.

Until now, extinction was part of nature

Endings are as important as beginnings. In geological time extinction certainly represents an ending but it also marks a beginning: death for one species but an opportunity for others. Extinction is a force that changes the world. Without it life would be impossible. The story of life on earth, through four billion years of upheaval and violent change, is one of constantly changing life forms, competing for resources, gradually increasing in diversity, and, it would seem, also in beauty and grace, and so onwards and upwards. Without extinction, there would be no forward motion; we would be stuck with never-ending, never-changing, primordial slime. In a state of nature, extinction (it is necessary to bear in mind), is positive, beneficial, life-affirming. *Homo sapiens*, for instance, exists on a mound of bodies, the trail of long-lost vertical apes and hairy ancestors.

One of the awakenings of the past half century is the realisation that we live in a violent universe. Our world circles the sun in a solar system of stunning turmoil. Bodies crash into one another, orbits spiral out of kilter, and explosions of rock hurtle out into space to eventually land as meteorites. A big meteorite or comet casually wiped out the dinosaurs and created a *tabula rasa* for our mammalian ancestors to exploit. On earth exploding volcanoes pour greenhouse gases into the atmosphere, and occasionally the eruptions are massive enough to disrupt life not only locally but across the world. Our planet has experienced cycles of torrid heat and bitter cold, creating deserts and ice caps, and wiping out whole families of species unable to adapt to the new circumstances. On at least five occasions in the past 500 million years three-quarters of all species on land and sea were wiped out. These events are now known as mass extinctions: the Big Five. The last one happened 67 million years ago. It wiped out nearly everything bigger than a cat.

We are now witnessing another mass extinction event. They call it the Sixth Extinction. The one we started all on our own.

The human volcano

Today, as all the world knows, we are in environmental crisis. It is caused, fundamentally, by the burning of fossil fuels: the human-industrial volcano, so to speak. Burning releases locked-up carbon in coal, wood and oil, which joins the atmosphere as carbon dioxide, a 'greenhouse gas'. Greenhouse gases help to seal in solar heat. They make the world warmer, and that, in a state of nature, is a very good thing. We would be a very cold planet without some carbon dioxide to blanket the world. Unfortunately an excess of greenhouse gases makes the world warmer still, uncomfortably warm. The atmosphere now has half as much carbon dioxide again as in pre-industrial times (412 parts per million compared with about 280). In these proportions, carbon dioxide can be a killer.★ It causes climate chaos, acidifies the oceans, expands the deserts and closes down the future. Our emissions, measured annually by climate stations across the globe, now amount to some 40 gigatonnes per year and rising (Friedlingstein 2020). A gigatonne is a thousand million metric tonnes – tonnes of gas, that is – a figure so dizzyingly vast that, like astronomical distances, the mind cannot well grasp it. If you piled up every human being on earth into a huge human mountain, well, 40 gigatonnes is roughly twice the weight of *that*. It weighs as much as three million jumbo jets. And the systems of the world cannot cope with it.

The results are familiar to all, hideously so to some. Climate change results in more floods, more wildfires, more storms. Plants die, snow and glaciers melt, sea levels rise, and instability increases – and instability in turn creates suffering. The effects seem to worsen year on year. The extremes of heat and drought witnessed in 2020 saw forest

---

★ It nearly killed me once. I accidentally breathed in a hot lungful of CO_2 in a fermentation vat at a Scottish distillery, blacked out and nearly toppled in

and bush fires blazing across the world, from the Amazon to Australia. That year the Australian bush fires alone released 400 million tonnes of carbon, more than the country's entire human-made emissions. Hot blazes not only consume natural habitats and their wildlife but release yet more carbon dioxide, and equally damaging quantities of methane and nitrous oxide. Carbon emissions in Brazil rose by 9.6 per cent in 2019, mainly from fires, and Brazil is only the sixth largest emitter of greenhouse gases in the world (the European Union, though smaller in size, emits four times as much).

Relatively small increases in global average temperatures have disproportionately large effects. In 2019 the average increase was just under a single degree Centigrade above the twentieth-century average, and about two degrees above that of the nineteenth century. It doesn't sound much, but we know the consequences. The reason it hasn't risen any higher yet is because the oceans act as a heat sink: fortunately for humankind, and for life in general, three-quarters of the world is ocean, and because of the capacity of water to absorb heat it takes a truly tremendous amount of energy to raise global temperatures. But, by 2030, on present trends, greenhouse gases will have generated enough heat to overcome this thermal brake and temperatures will then rise more quickly, maybe by several more degrees by the end of the century. At the same time, the rapidly melting – and burning – permafrost is releasing yet more stored carbon. If nothing is done to stop it, global warming will make life considerably less comfortable. Today's children may look back with longing to the times when the grass grew green and you could live in cities by the sea and sunbathe on the beach.

On top of the disruption to the natural cycles of the earth, you have to factor in the world's rising human population. Since 1970 – within the lifetimes of at least a third of us – our numbers have doubled while those of wild animals and birds have roughly halved: a synergy that suggests a link. At the beginning of the Industrial Revolution, in the mid-eighteenth century, there were fewer than a billion people on earth. Since then, the world population has grown eightfold to 7.9 billion, and on present trends that number will be nearly 10 billion by 2050, and upwards of 11 billion by the end of the century (Roser et al. 2019). The populations of twenty-five countries are expected to at least double between 2020 and 2050. Most of this generation will live out their lives in cities, vast conurbations spreading across the landscape, blotting out forests, marshes, coasts. Somehow these conurbations will have to be fed from an environment which is already stressed by drought, pollution, soil loss and ecosystem collapse.

Ten thousand years ago, when humans hunted wild animals with flint-tipped spears and arrows, wildlife outnumbered us by ninety-nine to one, and there was no domestic livestock at all. Today, if we weighed every person on earth, we would amount to nearly a third (32 per cent) of the mass of every living animal on land. The rest (67 per cent) would be made up mainly of our flocks and herds. Collectively wild animals represent only one per cent of life on land, and even that tiny proportion is falling. Population Matters, the charitable body which tries to persuade us, ever so politely, to have smaller families, tells us: 'As long as our numbers are growing, the value of every other action we take risks being cancelled out by the demands and needs of new people joining the population.' To put it more bluntly: more people means less wildlife.

Using the wonders of the internet, with the facts and figures of science technology at our fingertips, it is open to anyone to work out roughly what the future will look like if climate change continues at the present rate. A realistic scenario would need to take into account facts and trends, and probabilities based on them, and the likely demand on the earth's resources. What it is impossible to know is the human factor – whether we will respond intelligently or stupidly to the problems that face us. Certain scenarios will almost certainly not happen. It would be a good outcome if we were selfless enough to leave half the world's open spaces for wild animals and plants, as the late ecologist Edward O. Wilson advocates. But it isn't happening. Or we can imagine a world made climatically safe by a fast global reversion to carbon-zero as Extinction Rebellion and Green parties would wish us to do. That isn't very likely either – it might be achievable in particular countries but surely not soon and not worldwide. Perhaps we live on such illusions a bit too much. Humans thrive on hope, which elides into plans, declarations, resolutions. Of course we all have to hope, but, as sixteen-year-old activist Greta Thunberg had to remind us, hope without action is useless. And it needs to be the right action. Environment-friendly action might mean people-denying action. But how many of us would support a government that denied us our deepest desires, our aspiration for a better life?

But without such action, and meaningful sacrifice, it is inevitable that we are going to say goodbye to a lot of the animals and plants and insects with which we share the world.

At risk of extinction according to the IUCN's Red List of Endangered Species are

40% of the world's amphibians
34% of the world's conifers
33% of reef corals
31% of sharks and rays
27% of 'selected crustaceans'
25% of mammals
14% of birds

That is, if we carry on as we are now. If we do nothing to help. Some of these species are probably already gone.

A guide to what follows

The basic facts about climate change are becoming familiar, perhaps wearyingly so. But its consequences are usually discussed from a human-centred perspective, understandably enough. From the World Economic Forum to COP26 to purely local initiatives, the focus has been on green energy and carbon reduction for the sake of future prosperity. *Our* prosperity, that is.

The impact of climate change on the extinction rates of wild species is a secondary consideration, and usually talked of as though it were more of a threat than an actuality. But unfortunately the Sixth Extinction has already begun. The world's wildlife is in steep decline – numerically speaking, most of the world's rhinos and great apes, for example, have already gone – and hundreds of species have become extinct in the past fifty years. Thousands more are likely to follow in a process that is probably unstoppable. Ecosystem collapse and the extinction of species is, or should be, a defining theme of the present, not a projection of the future. To repeat, it is wrong to call mass extinction a threat. It is the inevitable accompaniment to the human-dominated, fast-changing world we live in. It cannot be a threat because it is happening already, right now.

This book represents my own take on a subject which is naturally depressing, but which, I think, is also extremely interesting. Why has extinction been the fate of some species but not others? How does extinction affect us as feeling and thinking human beings? The threat of extinction can bring out the best in us, in our determination to try and save some of the most attractive species on earth. It has already created an unparalleled burst of scientific research and survey. Half of the world's frogs, for example, were discovered and described in the past fifty years, and the same is true of the world's sharks (there is a strange synergy at work here because roughly half the world's frogs – though not necessarily the same ones – are also believed to be threatened with extinction). Conservation efforts, locally and worldwide have been spurred on by extinction. Many of the most threatened species on earth now have resourced action plans for recovery. Of course a plan for recovery is not the same thing as an actual recovery (though you sometimes are given the impression that it is). As Robert Burns reminded us, even the best laid schemes *gang aft a-gley*. And failure in this case is not something that can be corrected later. Nothing on earth is more final than extinction.

To prevent this book becoming a jeremiad, or a dreary listing of threatened and lost species, I have tried to approach our deathly subject thematically, in the diverse ways in which extinction happens, in the plants and animals it hits, the places where loss is at its greatest and in its impact on our sensibilities. First, I review the great mass extinctions of the past and their relevance to what is happening today – for if you want to fix something, first find out how it works. Then I review why extinctions happen in the modern world, and of some of the extraordinary efforts done to prevent it. We tend to think of threatened species as animals and birds – the ones that appear on television – but the greatest number is among those small beings that run the world, the invertebrates. It is so easy to forget them. After this worldwide review, I turn to the lands I know best: Great Britain, that is, England, Scotland and Wales, where the slow bleed of biodiversity is linked to habitat fragmentation: the rags of wild nature surviving in an overcrowded land.

In Chapter 4, I address the core of the issue: how to survive and avoid the queue that leads to the door marked exit. It addresses an imaginary species in terms of dos and don'ts. Some things you should avoid at all costs, while certain survival skills will give you a better chance. I hope this chapter will not seem facetious or uncaring, but it seemed to me a necessary way of making a lot of diverse information readable and even, dare I claim, amusing. We then move on to pathways of extinction: the journey of a species to its grave. Every lost species has its own story, of course, but I think my chosen five, a fish, a moth, a dolphin, a wolf and an earwig, collectively say something about human perception of loss as well as about extinction. In Chapter 6, we are reminded that extinction isn't always the last word, for a species may have a posthumous life in our imaginations and our products, as icons and symbols, even as film stars. That is why I feel able to include everybody's favourite dinosaur, Tyrannosaurus rex, which hasn't been around in the flesh for a very long time (and never co-existed with humans), and one species – the dragon – that never existed at all.

Chapter 7 is about our reaction to extinctions, as the actuality of the Sixth Extinction dawns on us, the first mass extinction of life since *T. rex* perished in a fireball. Rebellion is in the air, but on whose behalf, our own environment or the natural world? The last chapter pursues reasons to be more hopeful, ending with the consoling delights of nature, extinction-bound or not. Overseeing all is the crucial matter of tone. I find pure science writing colourless because it forbids normal human emotion. Similarly the moral earnestness that pervades much of conservation writing can become wearisome, as well as jargon-infested. It seems to me that we don't have to write as though we were making a sermon. Admitting that in the matter of extinction of life we are all sinners by default does not preclude the possibility of irony, sarcasm and nervous laughter (laughter in the dark). You can't be detached about extinction. And, when looked at in a certain way, although extinction is certainly tragic, it contains other facets too, disgust, irony, fascination (the fascination of a rabbit in the headlights), even comedy. It doesn't pay to be serious all the time, even about a subject like extinction.

There is extinction, and there again there are extinctions, local, national, worldwide; extinct in the wild, possibly extinct, probably extinct, and 'Data Deficient', meaning that, yes, you are almost certainly extinct. There is extinction by neglect, and extinction by deliberate act, and 'declared' extinctions that turned out to be wrong and other newly declared extinctions that actually happened a long time ago. There is even life after extinction, a kind of cultural after-glow. Extinction has many meanings and many possibilities as a subject. That is why I have subtitled this book *Extinctions*, and not *Extinction*. There are many ways in which life comes to an end, a fate with many colours.

Perhaps I had better say a little about myself. I am not a professional scientist, though I have of course studied science, at school, at university, and also didactically. Rather I am an old-fashioned, all-round naturalist. I have been a professional nature conservationist, working in Scotland and England. I have co-led wildlife-focused tours around Europe and travelled parts of the world in search of wildlife, as well as nature forays at home. I have written about wildlife and naturalists and nature conservation across two dozen books and countless feature articles and reviews, both in the specialist and also the national press. I have observed the conservation scene over half a century and taken a minor role on that great stage, the equivalent I suppose of the assistant gravedigger in *Hamlet*. I have saved this book until my eighth decade. Extinction is nature's last word. Whether it will be my last word too, I don't know. You never know do you? Hopefully not.

Chapter 1

Extinctions through ancient time

> Life is a copiously branching bush, continually pruned
> by the grim reaper of extinction, not a ladder of
> predictable progress.
>
> Stephen J. Gould (1989), *Wonderful Life.*

What was the first species ever to die out, the very first
lost life on earth (and so, you might say, the least successful
species ever)? It's hard to say. For the first three billion
years, life on earth consisted solely of micro-organisms,
visible only as slime or coloured hazes in the water or
strange projections from the shallow seabed topped by a
greenish crust. Did some forms of slime do better than
others? Was extinction a natural phenomenon from the
very beginning? Until recently, it was believed that
micro-organisms like bacteria or viruses were more or less
extinction-proof. They exist in such unfathomable numbers,
like sand grains on a beach, that nothing short of a universal
deluge would ever stop them in their tracks. *We* can kill
off a micro-organism, given a worldwide vaccination
programme, such as the smallpox virus, or one of the
bacteria that causes leprosy. But does nature? Fossils in this
case do not help much. It is hard enough identifying living

micro-organisms, let alone billion-year-old mineralised specks in the rock.

Recent research by Stilianos Louca and colleagues at the University of British Columbia (2019), Canada, took an indirect approach to the problem. Rather than studying fossils in situ they used 'massive' DNA sequencing, followed by equally massive data analysis to construct a family tree of bacteria, showing how the microbes have diversified over time. On that basis, Louca estimates that there are between 1.4 and 1.9 million 'lineages' of bacteria in the world today, many of which have yet to be discovered and named. The same data indicates that between 45,000 and 95,000 kinds of bacteria are likely to have died out during the past million years alone. In other words, species of bacteria become extinct in the end just like any other form of life. Within their own micro-worlds, bacteria compete for resources and the strong species survive while the less well-endowed go under. No forms of life, even the most primitive or the most numerous, are proof against the ravages of time, change and competition.

So the first extinctions on earth were almost certainly micro-organisms that we know nothing about and wouldn't have noticed even had we been around at the time. My own candidate would be an Archaean ('ancient thing'), an antediluvian micro-organism that derives its energy from the oxidation of methane. Imaged through an electron microscope they look like jellybeans. Archaeans still exist, although nowadays they might be happier on another planet. The yellow colouration gathered around the hot springs of Yellowstone Park is made up of a universe of Archaeans (though unfortunately it is not the 'yellow stone' of

Yellowstone; that is the name the native inhabitants gave to the rocks along the river). Perhaps the Archaeans were the first bright colours on earth, the living froth around the vents of geysers. Maybe the very first extinct being was all the colours of the rainbow. Nature's show-offs rarely succeed for long.

Extinction is the great decider. It deletes the less fitted – which were of course perfectly well fitted in their day, until their environment changed – while sparing others, the ones that will go on and create new branches on the tree of life. At any point in time it would be impossible to pick out winners or losers. Evolution's great successes, such as horses or elephants, started out in an extremely modest way. The late, great, Stephen Jay Gould drew attention to the unpredictability of life's lottery: the 'billions of possible scenarios' (Gould 1989), potential outcomes from which only one will make it through. *Homo sapiens* is, in evolutionary terms, the merest twig on just one branch of a monstrous tree. If evolution had taken a different path, perhaps we might have ended up not with big brains and an aggressive attitude but as social innocents with a vegetarian diet living quietly in the woods. And, if so, the world might still be roamed by mammoths and dodos and sabre-toothed cats. Like all species, we owe our existence to a coincidence of time, circumstance and evolutionary opportunity.

When I studied fossils at university, long ago, I remember Robin Wootton, our amiable lecturer, introducing us to *Tullimonstrum*, the Tully Monster, named after a fossil collector called Francis Tully. This creature, only the size of a prawn, swam in ancient seas at about the time when forests were first forming on land. The point about

Tullimonstrum is that, despite a fair amount of detail on the fossils, no one can work it out. This little monster is roughly fish-shaped but has a long proboscis sticking out from the front like the hose of a vacuum cleaner, ending with a pair of snapping jaws each bearing six sharp teeth.

'You are all quite experienced in animal anatomy by now,' I remember Robin saying. 'So where would you place this?'

Some think *Tullimonstrum* is a kind of worm, others a sort of shrimp, while the majority plump for an early bash at a vertebrate, perhaps a distant relative of living lampreys. If the latter guess is right, *Tullimonstrum* might have become a vertebrate ancestor, giving rise to whole families of funny little fish shaped like vacuum-cleaners. As it was, *Tullimonstrum* was on the wrong team. It soon dropped out of the fossil record, to surface again only 280 million years later as the state fossil of Illinois.

Another might-have-been was *Ichthyostega* ('fish-roof'), an ancestral amphibian which looked a bit like a giant tadpole with feet. What was unusual about *Ichthyostega* was its hind feet, each of which bore not five toes, the accepted base number for all land vertebrates, but seven. Yet the idea of seven-toed animals never took off, and we can only imagine what the history of music, say, might have been like if *Ichthyostega* had been our ancestor and we all had seven digits on the end of our arms instead of five.

Surely everyone who loves prehistoric life has a favourite animal? My own top beast is a mammal, the first really big mammal, called an uintathere. It was there right at the start, in my toddler's colouring book, a big, snorting, frightening-looking beast that, I recall, was looking me squarely in the eye. Actually there were two of them, *Uintatherium* ('Uinta beast') and the even bigger *Eobasileus* ('dawn emperor'). Both had huge heads with no fewer than six horns, arranged in pairs along their low, flat skulls, plus a pair of long, curved fangs that wouldn't have been out of place on a sabre-toothed tiger. Their heavy bodies were wide-hipped, like a hippo's, but with longer, elephantine legs ending in a semicircle of five hooves. So they looked a bit like a rhino, and a bit like a hippo, and a bit like an elephant, and since some evolutionists thought they came from the same unlikely lineage as lagomorphs, maybe with a bit of rabbit thrown in too. In his reconstruction of the beast, the nineteenth-century fossil hunter, Edward Drinker Cope even gave them an elephant's trunk and a pair of huge flapping ears (reasoning that if they were the size of an elephant they probably looked like an elephant). The Natural History Museum once had a whole gallery devoted to fossil mammals, and among them was a skeleton of *Uintatherium* standing proud in front of a painted backdrop of tropical palms, an awe-inspiring sight for an open-mouthed boy such as myself. But that gallery has long since been dismantled and my favourite beast despatched to some basement. Which is why children today have never heard of the great beast from Uinta (and neither has Microsoft's spellcheck).

The uintatheres did not last very long. They were the giants of the mid-Eocene, about 40 million years ago. They 'ruled the earth' for a while but died without issue. They left no descendants and today, as with the Tully monster, we cannot even agree on what kind of animals they were. Perhaps what weighed against their survival was the fact that they were incredibly dim. Their huge bodies were controlled by tiny brains, sufficient, obviously, for their needs, but perhaps inadequate once there was competition from slightly brainier animals. I still dream of a world where uintatheres survived and gave rise to whole dynasties of grotesquely horned and fanged beasts, bellowing from their primeval swamps as the sun dips behind the peak in a scream of golden light.

This is not to say there was anything inherently wrong with uintatheres, or Tully monsters for that matter. They did well enough to enter the fossil record, and perhaps any evolutionary biologist transported back to their time might have seen them as very promising lines of development. It is, in fact, impossible to spot unfitness in the fossil record. As the American palaeo-biologist David Raup (1992) noted, the only hard evidence for unfitness is extinction, and in that sense every species is eventually unfit. Survival seems to be as much a matter of good luck as good genes. The uintatheres shared their humid forests with the first horses, nondescript little animals the size of terriers. Yet the latter's descendants spread all over the inhabitable globe while the uintatheres soon vanished forever. Were their contrasting fates just a matter of luck? And if so, is there not something a little fearful about the random nature of events? You can't bet on being a survivor. All you can do is hope.

Mass extinctions

In the books of prehistoric life that I grew up with, extinction was usually thought of as a gradual business, a slow weeding out in an orderly Darwinian progression. Admittedly there were sudden unexplained crashes, such as the moment when all the dinosaurs disappeared, apparently quite

suddenly, at the end of the Mesozoic. That particular extinction was often ascribed to the dinosaur's basic, cold-blooded
stupidity and the inevitability of a takeover by the brainier
mammals (unless you follow Will Cuppy's satirical view
(1941) that 'the Age of Reptiles ended because it had gone
on long enough, and it was all a mistake in the first place').
But as scientists looked more closely at the fossil record,
this idea of gradual change didn't seem to stand up.

In 1972, Stephen Jay Gould and Niles Eldredge came up
with the alternative theory of 'punctuated equilibrium'
(Eldredge & Gould 1972). The record suggests that far from
being orderly and progressive, life goes on without much
change for millions of years, cruising along thriftily, until
suddenly, bang, an environmental catastrophe intervenes
and everything collapses in chaos and death. Then things
start over, and the survivors inherit the earth as a land of
almost unlimited opportunity. Somebody summed it up as
'evolution by jerks'. Since then, Gould and Eldredge's
hypothesis has become more nuanced, with scientists like
Richard Dawkins arguing more for a process of stately
unfolding than a series of sudden jolts and jerks. The debate
swings to and fro and goes on. Partly it depends on how
you define a 'jerk'. In geological terms, it could last tens
of thousands of years but it still looks as though evolution
had been kicked in the pants.

What is beyond question is that there have been times
in the earth's history when the majority of life did die out
suddenly in a single convulsive event (or at least a succession of such events). These very occasional catastrophes are
called mass extinctions. As the British palaeontologist
Michael Benton put it, think of the tree of life, and then
imagine 'vast swathes of the tree are cut short, as if attacked

by a crazed, axe-wielding madman' (Benton 2015). That is a mass extinction. But afterwards, just as trees will often grow up again from a cut stump, so new life will spring up from the limited pool of survivors. The pack is reshuffled and life goes on: a disaster for some, in fact for most, but an opportunity for others.

There have been at least five such moments when three-quarters of all plant and animal life died out, and when the bountiful earth briefly turned into hell. Ninety-nine out of every hundred species are extinct but sixty-five of those were lost during the Big Five mass extinctions. The earliest of these took place at the end of the Ordovician, 444 million years ago (although it was preceded by what was at least a major extinction event a hundred million years earlier, at the end of the Ediacaran period (*q.v.*)). It happened again at the end of the Devonian, 359 million years ago; at the end of the Permian, 252 million years ago; at the end of the Triassic, 201 million years ago and, for the fifth time, at the end of the Cretaceous, 67 million years ago, the famous K/T event which saw off the last dinosaurs.

The regularity of these events – you could almost call it a pulse – is assumed to be a coincidence. Each one was the result of a separate set of circumstances and were natural events. Nonetheless, they have relevance to our own situation, here and now, for the radical changes that brought about extinction in the deep past are now being replicated by humankind. As Elizabeth Kolbert put it, 'we are running geologic history not only in reverse but at warp speed'.★

★ *The Sixth Extinction: An Unnatural History* (Henry Holt and Company: 2014)

The sixth mass extinction has started. What will survive this time around?

I think it is worth revisiting these ancient catastrophes for the light they shed on the present. But let's begin with a mass extinction that is not yet on the official list: the untimely end of the so-called Ediacaran Biota: life's false dawn.

Killing times

Prologue: the Ediacaran, 540 million years ago

In 1957, a schoolboy messing about in a quarry in Charnwood Forest, Leicestershire, spotted a strange object embedded in the rock. It was shaped like a leaf but with the padded texture of a quilt and was obviously a fossil. That was surprising because these rocks were of Precambrian age, a remote time when there were not supposed to be any fossils, or at least not large ones, and therefore no life, apart from slime. As every schoolboy knew, proper fossils arrived with the Cambrian. The boy showed it to a geology lecturer at Leicester University, who wrote a scientific paper about the find, naming the object *Charnia* in honour of Charnwood Forest, and *masoni* in honour of the boy whose name was Roger Mason.

That schoolboy could have been me. Ten years later, I too was looking at life in the disused quarries of Charnwood Forest, after bunking off games at my school. But I was studying pond life, not the world's oldest fossils, of whose existence I had no idea. Evidently the then youthful Sir David Attenborough, once a keen young fossil hunter from Leicester, has had the same regret. I was too late, he too early; what a shame, it could have been a big break for either of us. Today, these ancient fossils are famous and, in some places in Charnwood Forest, you can walk right up to them and touch their faint imprints in the rock. One of the most accessible is now covered with graffiti, a crisp enough comment, I feel, on our times and the reverence we feel for *Charnia masoni*.

Pre-Cambrian life was first revealed to the world in rocks from the old mining country of the Ediacara Hills in South Australia. Hence the fossils of this time are now known as the Ediacaran Biota ('the life of Ediacara'). Similar fossils have since been found in sedimentary rocks all over the world, and some 100 genera of life forms have been described, ranging from leaf-like objects like *Charnia* to grooved lumps resembling giant thumbprints or flattened footballs. In reconstructions of that primeval seabed *Charnia* has been interpreted as a sea-pen, waving gently in the water, in various pretty pastel colours (though why they have colours when there were no eyes to see them is hard to explain). Actually it is not a sea-pen at all but a deeply archaic form of life with no modern descendants.

Many of these creatures are hard to fit into the living world we know today. Their evident lack of legs or jaws, or even heads, suggests that life on the seabed between 600 and 540 million years ago was crowded with soft-bodied blobs or gentle filter-feeding animals that may have looked more like flowers. It has been called the 'Garden of Ediacara', perhaps a time of peace and tranquillity never to be seen again in the world. In 2004, the Ediacaran became the first new geological period created for more than a century.

Few of these enigmatic creatures survived long into the Cambrian. Their loss could be called the first mass extinction.* What presumably did for them was the evolution of faster moving, more advanced animals with hard body parts, including jaws. They call this extinction by 'biological interaction', which is a polite way of saying that the Ediacarans were torn to shreds and devoured. The Pre-Cambrian garden was replaced by the far busier world of the Cambrian, a sea world of shells and shrimps, a place of suddenly 'unbounded ecological opportunity' (Leakey 1996). In place of meek, slow-moving blobs there were now 'crawlers, walkers, burrowers, slurpers', biters. It seems that oxygen in the atmosphere, and in dissolved form in the water, had by then increased to the point where animals could metabolise collagen, the protein building block for hard body structures including an external skeleton. Life became more competitive: an eternal contest. As aggressive, competitive animals ourselves we are bound to see the Ediacaran extinction as progressive and highly necessary. Come on, evolution, we might say, let's do it! The gloves are off. The arms race has begun. (Can't wait for the first sharks!)

All out in the Ordovician, 444 million years ago

The Ordovician, the second division of the Palaeozoic ('ancient life') era, lasted from 485 to 443 million years ago. Like the Cambrian and the Silurian on either side of it, the Ordovician was named after Wales. Pioneer geologists

---

* There was an episode of 'snowball earth' about 653 million years ago, when sea ice extended to the equator, and presumably caused widespread extinction, though the fossil record of that period is little known.

spent a lot of time in Wales, probably on slippery rocks in the pouring rain, and they gratefully named the strata after ancient Welsh tribes, the Ordovices and the Silures. The Cambrian was named after Wales itself, in its Latin guise of Cambria.

During the Ordovician, most of the land lay in the southern hemisphere. The rest was ocean. The land was still barren except along the coastal fringe; there were no fully terrestrial animals, but the shallow seas teemed with sponges, scuttling trilobites, primitive corals and sea lilies, strange, stalked creatures such as cystoids and blastoids, and, above all, the world's first cephalopods – tentacled animals with cone-shaped or coiled shells known as nautiloids. The smaller nautiloids were able to move up and down in the water column and so inhabit the open ocean as well as the seabed. Life went on in those placid warm seas, year on year, millennium after millennium, until, towards the end of the period, something awful happened.

For some reason the planet suddenly turned much colder. Ice sheets replaced the warm southern seas and there may even have been a second period of 'snowball earth' when polar ice reached towards the tropics. As water turned to ice, the sea level fell and former shallow beds along the continental shelf turned into arid plains. At the same time, dissolved oxygen levels fell and the hitherto clean, bright sea water turned turbid and even polluted. Why this happened is uncertain. There is no sign of major volcanic activity at the time, or of the earth being hit by a meteorite (which isn't to say they didn't happen). It may be that the uplift of mountains, such as the Appalachians, triggered the large-scale weathering of rocks and the consequent absorption of carbon dioxide from the atmosphere. However it may have been, the event created sudden, drastic cooling and destroyed life on an unprecedented scale.

At least half the main groups of animal life, both genera and families, disappeared, including most of those splendid nautiloids, which included some of the first really large animals on earth (all right, some of the previous Cambrian Dinocarids or 'terror crabs' were pretty fearsome-looking). In the mind's eye one can picture barren wastes of cold mud under ominous clouds, made more eerie still by the projecting spires of empty nautiloid shells. Everything looks dead. In fact nearly everything *was* dead.

If you count the end of the Ediacarans, this was the second mass extinction in the earth's history, and possibly the second biggest. In terms of the proportion of life lost, it was even more devastating than the famous K/T asteroid impact that wiped out the dinosaurs. So why do we not hear more of the great Ordovician extinction? Probably because the replacement life was, at least to a non-expert eye, pretty much the same as before. The cold wiped out trilobites, brachiopods, conodonts and primitive corals. And they were replaced by . . . more trilobites, brachiopods, conodonts and slightly more advanced corals. The planet, as it were, shook itself as if after a long sleep; the ice melted, the temperature rose and life resumed.

Devonian disaster, 359 million years ago

I wonder how many shoppers at Peterborough's Queensgate shopping centre realise that they are stepping over an ancient reef. When this place first opened, in the mid-1980s, I remember spotting chambered spiral shapes in the dark matrix of the slabs, along with various fuzzy blotches and bits of what were plainly shells. The polished limestone had been cut from a quarry in Germany whose rocks came from what had once been a coral reef. The Queensgate slabs are later in age than the Devonian, but the fossils are of a broadly similar type. The Devonian, named after the limestone rocks of Tor Bay and often known as the age of fishes, was also an age of coral, formed in warm, clear, shallow seas which then stretched almost from pole to pole. Devonian corals were rough-surfaced ('rugose'), horn-shaped objects from which emerged polyps something like a sea anemone – flowers of the reef. Other types were colonial, resembling a honeycomb. At university, I remember sitting through whole lectures on these corals (waiting impatiently to get on to more interesting stuff). Evidently some of them, although incredibly abundant, were quite short-lived as species, and so they are useful for dating rocks precisely, as a kind of coral clock. Hence knowing about ancient coral comes in handy when, for example, looking for suitable slabs to pave a shopping mall.

As before, it all went tumbling down in the end. The end-Devonian was a time of shifting continental plates, with much volcanic activity along the margins, both on land and under the sea. It was also a time of abrupt climate change with extremes of heat and cold. But a new problem emerged in the unexpected form of 'killer trees'. Plants were now colonising the land and they drew carbon dioxide from the atmosphere, grounding it as plant tissue. In the process they created the world's first soil, rich in nitrogen and phosphorus. Soil is obviously good for plants, but when it is removed and transported to the sea as sediment in rivers, it effectively fertilises the oceans. This was something new. Fertilised water produces blooms of plankton which also remove carbon from the atmosphere and in addition cause anoxia, the removal of oxygen from the water. Without oxygen, animal life, such as fish, will die (as we see regularly on de-oxygenated canals and lakes). In some rocks this period is marked by a layer of death: the black shales. Hit the soft rock with your hammer and you may even get a whiff of petrol: the preserved stink of the dying ocean 360 million years ago.

The end of the Devonian might have been the great age of mud. Mud is good for plants but bad for coral. And what is bad for coral is bad for fish. Whole ecosystems crashed, and great swathes of life abruptly disappear from the fossil record: whole families of brachiopods and other shellfish, and nearly all the armoured and lobe-finned fish that defined the Devonian age. Some 97 per cent of vertebrate life in the sea disappeared. The surviving fish were all small, for, in times of strife, small life usually does best. The demise of giant armoured predators like *Dunkleosteus*, the great white shark of its day, paved the way for the dominance of sharks as the apex predators of the seas, a role they have enjoyed ever since.

The Devonian disaster poses a warning of what can happen when the balance of the earth suddenly tilts. Today the coral reefs are dying once again. Half the Great Barrier Reef has gone in recent years. I have swum over coral reefs in three continents and nearly every one of them was bleached and half-dead. Reefs are among the most bio-diverse habitats on earth: the 'rainforests of the seas'. But coral exists on a climatic knife edge. It thrives only in clear, warm water within a fairly narrow temperature range. Today it is global warming that is bleaching and killing the reefs; back in the Devonian, it seems to have been global cooling. Some of those ancient reefs didn't recover for millions of years. We probably don't have that long.

The Carboniferous coal crash, 299 million years ago

The Carboniferous is named after coal (it means 'coal-bearing'), and coal is the petrified remains of ancient plants. It did not end with a mass extinction, like those which preceded and followed it, but all the same life did change quite radically as the period drew to a close. And this time we have a good idea why.

The Carboniferous was unique in the earth's history because it had tall, dense forests but no really large land animals. Instead some of the newly evolved land invertebrates grew to an impressive size – for invertebrates – with primitive dragonflies the size of seagulls and fearsome two-foot-long scorpions. And so the trees were churning out oxygen as a by-product of photosynthesis without any big beasts to breathe it in. The consequence was the highest ever proportion of oxygen in the atmosphere – perhaps as much as 35 per cent compared with 21 per cent today – producing a very flammable form of air in which forest fires must have been commonplace. The age of coal could equally well be called the age of oxygen. At the same time, and for the same reason, carbon dioxide levels in the atmosphere fell, resulting in cooling, and so what started as a kind of global rainforest ended up with a more temperate climate. In place of fire came ice. Glaciers formed at high altitudes and at the poles, and those lush swampy forests began to recede. The crux seems to have been this imbalance of plant and animal life. The plants created changes in the atmosphere that led to their undoing. The amphibians and giant invertebrates that lived in those forests needed to adapt or die; and a lot of them died. Interestingly the most abundant insect survivors were cockroach-like species.

Cockroaches are famously adaptable insects, as much at home in a dirty kitchen today as they were the debris of the coal forests.

In a sense the forests did not disappear. Some of their carbon, and their energy, is preserved as coal. The reason why this age, but not others, produced so much coal is uncertain. Possibly there was less decomposition because organisms had not yet evolved that could digest cellulose and lignin, the hard materials that make up wood. If so, the dead wood will simply have piled up, unless consumed by invertebrates or burned by wildfires, to be compressed into peat and then coal. That accumulation of carbon in un-rotted debris was another reason why the Carboniferous cooled down: it reduced the amount of carbon dioxide in the atmosphere. Eventually humankind unlocked this carbon by mining and then burning the coal, thereby thermally decomposing the mineralised wood and releasing that long-stored carbon. And, as we now know, in so doing we caused the climate to change, as it did with all the mass extinctions on earth. Effectively we burned the Carboniferous.

It might have been a different story had fungi been there to rot down the plant remains. Today fungal decomposition is one of the largest sources of carbon emissions at around 85 gigatonnes of carbon per year (burning fossil fuels accounts for only one-eighth of that). Assuming that fungi were not sufficiently advanced to play this role in the Carboniferous, the carbon dioxide removed from the air by plants would not be replenished. And so, without fungi to help them, those ancient plants were ultimately the authors of their own demise.

Hell on earth: the end of the Palaeozoic, 252 million years ago

There was a moment in the world's history, just one, when circumstances came close to wiping out all life on earth, on land, in the air and in the water. It demonstrated that, under certain conditions, our beautiful blue planet can turn into an arid, stinking, baking-hot hellhole – and, it seems, with disconcerting speed. At around 252 million years ago, the Palaeozoic, that long reign of ancient life, came to such an end.

We know that it happened and what the consequences were. The hard part is to determine exactly why. What caused it? The end of the Permian is difficult to study because there are so few rock formations of the right age (or rather, ages for there were a number of extinction 'pulses' before and after the big one). The key to the disaster seems to have been the displacement of the continents at that time. By the Permian, continental drift had resulted in the collision of different land masses to form a single huge continent known as Pangaea. It straddled the equator and extended almost to the poles, surrounded by a global ocean. This configuration meant that much of the land interior was desert – a barren land stained red, like Mars, by oxidised iron, which is still visible in the rocks of the New Red Sandstone, laid down during this period. It also produced much volcanic activity, notably a vast spewing of lava known as the Siberia Traps, accompanied by a big increase in greenhouse gases. The main one was carbon dioxide but there was also a lot of sulphur dioxide which, combining with water in the clouds, fell to earth as acid rain. At the same time, atmospheric oxygen became depleted as a result of the decline of green plants and ocean plankton.

Locked into a spiral of destruction, the world grew hotter and hotter, by as much as five degrees Centigrade. But even that wasn't the end.

Even such hot, breathless conditions would not necessarily account for a mass extinction at this point. The world was now dominated by reptiles, the first fully terrestrial animals, and by plants such as seed ferns and early conifers which were better adapted to a dry world than the giant clubmosses and horsetails of the Carboniferous. Some of the reptiles were taking on mammal-like attributes such as upright legs, specialised teeth, a herbivorous diet and, in certain cases, perhaps even bristles and whiskers. Evolution was taking life in a new and interesting direction. But instead of producing the world's first hairy, warm-blooded animals, life fell off the cliff. If our estimates are even reasonably accurate, comments David Raup (1992), 'biology had an extremely close brush with total destruction' (and it does make you wonder whether there are other earth-like planets out there which were not so lucky and that nothing at all survived).

Why was this mass extinction the worst? One possibility is ocean burping. Shallow seas became swamped with debris from the land, and parts of the deep sea turned anoxic, that is, their dissolved oxygen became depleted. In these conditions, bacteria that derive their energy from sulphur compounds will thrive in the deep sea. Their respiratory product is not oxygen but hydrogen sulphide – a smelly, corrosive gas that would have bubbled up from the ocean depths and onwards, high up into the atmosphere, where it would promptly attack the ozone layer that protects the world from harmful radiation. With the ozone layer depleted, ultraviolet radiation from the sun would pour down unimpeded, causing all manner of ulcers, cancers and immune system damage, ultimately bumping off all those promising mammal-like reptiles. Effectively, they would have died of sunstroke.

Another possibility is that ocean temperatures increased to the point where quantities of methane, normally bound in the crystal lattices of ice under the ocean floor, were released, bubbling up into the atmosphere, adding to the already serious greenhouse effect and increasing global temperatures by *another* five degrees or so. An extra ten degrees in an already warm period would have made life very uncomfortable, perhaps destroying vegetation except at altitudes or along the coast. Finally, there is always the possibility of an extra-terrestrial strike or even a nearby supernova, drenching the planet in lethal radiation. A 500-kilometre crater discovered buried beneath the rock and ice in Antarctica would fit the bill, although no rock samples are yet available to test this possibility.

Can you imagine it? A dead sea turned acidic and belching poisonous gas, the shoreline blackened and stinking of death, and above, heavy clouds, grey and yellow, parting to reveal a sickly green sky. Most of life on earth seems to have died out within 100,000 years or less, an eye-blink in geological time. In the sea, a staggering 96 per cent of all species perished, including the last trilobites, a group that had flourished for hundreds of millions of years. On land, 70 per cent or more of all species were lost, including the first and only known mass extinction of insects. With the further pulses of volcanic activity that followed the catastrophe, ecosystems did not stabilise for at least 5 million years, perhaps not fully for 30 million years – and, if so, just in time to crash again. The survivors of the wreck staggered out into a ruined world, all harsh rock, desert and scrub, the unpromising opening of the Mesozoic ('middle life').

What did survive, rather surprisingly, was a buck-toothed, pig-like reptile called *Lystrosaurus*. Whatever its secret – maybe it lived deep in a burrow on the riverbank – it now had the world more or less to itself. For a few million years it was perhaps the most abundant large animal of all time. The equivalent survivor in the oceans was a scallop-like mollusc called *Claraia*, which became 'pandemic' on the desolate seabed, grazing the remains of the infinite dead. Perhaps the most long-lasting consequence of the Great Dying is that it put back the evolution of mammals, or at any rate mammals bigger than a rat, for the duration. The next stage in the world's history did not belong to them but to more reptile-like beings, leading, eventually, to the long reign of the dinosaurs.

The Permian disaster forms an object lesson in how to kill everything: how to mangle and crush all life from the deep ocean to the beach, to the lakes and rivers, to the forests and on to the mountains and deserts inland. It reminds us that life on earth is contingent. Life depends on the world's systems and cycles – carbon and nitrogen especially – working in balance. Thrown out of kilter, chaos descends.

Relatively small changes in the atmosphere, caused by the build-up of greenhouse gases, have disproportionately far-reaching consequences. In normal times, the Gaia effect, the self-limiting systems of the planet, work well enough. But push things too hard and the balance tilts. It takes us by surprise. The result is a runaway cycle of destruction, worse news piled on bad, until the whole steaming, poisoned mess, all its rotting vegetation and bleached bones, lies lifeless beneath the burning sun. At the end of the Permian the process may have been driven by volcanos. But as Sir David Attenborough put it in his 2021 television show *A Perfect Planet*, *we* are the volcano now. A big, roaring, heedless, super-volcano, super-heating our world.

The Triassic turn-over, 201 million years ago

The Triassic opened in carnage and ended in carnage. The whole epoch represented life trying again, radiating wonderfully, and then, abruptly, hitting the buffers once more. For the fourth time in the history of the earth around three-quarters of all species were wiped out. Although its effects were devastating and long-lasting, this latest swing of the scythe seems to have affected animals more than plants. In the oceans, the reefs, built up almost from scratch since the Great Dying, were destroyed again, along with their constituent corals, sponges, brachiopods and cephalopods. It marked the end of the conodonts, a long-lived group of eel-like animals, formerly known only by their teeth (technically their 'filter-feeding apparatus'). On land, a swathe of new, more advanced reptiles disappeared, among them mammal-like cynodonts, armoured aetosaurs and crocodile-like phytosaurs, plus some nightmarish crocodile-like amphibians – effectively giant meat-eating frogs. With all those potential rivals gone, the way was clear for the rapid advance and diversification of the dinosaurs in the succeeding period, the Jurassic.

The main cause of the Triassic mass extinction seems to have been the rifting of the super-continent of Pangaea. Just as the original coming together, the collision of land masses, was accompanied by intense volcanism, so too the subsequent rift brought with it increased carbon dioxide, global warming and the acidification of the oceans. There was also at least one sizable meteoric impact, whose crater rim is now the Manicouagan reservoir in Quebec, Canada. It was certainly big enough to wipe out life in the immediate area but seems to have pre-dated the main mass extinction.

And so out went another raft of wonderful animals and the ecosystems that supported them. And, once the world had recovered from that eco-disaster, in came a slightly more familiar world, the Jurassic Park of cycads, gingkoes and conifers with intervening plains of ferns, pterosaurs in the air and ichthyosaurs patrolling the oceans like dolphins, while the earth thundered to the tread of forty-ton sauropods and roared to the giant theropods plodding after them. Without the mass extinctions of their reptilian rivals, there might have been fewer dinosaurs, or even no dinosaurs. And without them, perhaps humankind might have evolved from ape-like ancestors a lot earlier than we did. In which case the Sixth Extinction might have happened, say, twenty million years ago, and I would be writing this in an unimaginable new world, after life, once more, was forced to start again.

The K/T event, 67 million years ago

One of the first films I ever saw was Walt Disney's *Fantasia*. There was a brilliant bit in the middle in which the story of the earth was choreographed to the music of Stravinsky's *Rite of Spring*. After a battle in which a terrified Stegosaurus is pulped by *T. rex*, the dinosaurs' lush world dries out and the parched animals march off to their doom, timpani drumming as they die one by one. Harsh stuff for a children's movie, I thought. Disney's man claimed it was a 'coldly accurate' account of the end of the Mesozoic, of the environmental change that ended the world of cerebrally challenged giants and prepared it for the advance of the mammals – and, ultimately, us. Up to 1980, it was assumed that the Big K/T, the Cretaceous/Tertiary mass extinction was much like the others, the result of climate change and a tougher environment ushered in by eruptions such as the Deccan Traps, which splurged red-hot lava over what is now India. Perhaps more than the other mass extinctions, this one seemed pre-ordained. The dinosaurs were surely too huge and backward to cope with change.

Then, in 1980, physicist Luis Alvarez and his colleagues proposed a different scenario. The boundary of the Cretaceous is marked worldwide, very conveniently, by a narrow strip of clay. Analysing samples of this substance, Alvarez found unusually high levels of iridium in it. This element, a precious metal related to platinum, is extremely rare on earth but comparatively common in rocky meteorites. That was suggestive. Alvarez also found tektites, tiny glassy beads that form after molten rock is sprayed into the air. Alvarez deduced from this evidence that the earth had been hit by a space rock roughly 15 kilometres across. The effect, he calculated, would have been devastating. The impact would have generated enough dust and noxious aerosols to cloak the entire planet and so usher in a dark, unseasonal winter lasting perhaps twenty years, long enough for every large animal on earth to starve to death, assuming they had not already been killed by the impact. Under these conditions plants that depend on sunlight, including plankton in the sea, would also die out. No other mechanism is needed to account for the loss of 75 per cent of all species, including all the dinosaurs and pterosaurs, all their monstrous cousins in the sea, and even most of the mammals and birds. A meteorite of that size striking the world amidships would manage it all on its own. And, as we know, the solar system is full of rocks that size.

The clincher was the discovery of a buried crater of the right size and right age: the 180-kilometre Chicxulub crater in the Yucatan peninsula in Mexico, one of the largest impact craters on earth. It is egg-shaped, indicating that the meteorite had struck at a shallow angle of 20–30 degrees, splattering debris in a north-westerly direction with the force of a hundred million megatons. Plant fossils, including water lilies, recovered from the debris suggest that the strike happened in midsummer. In 2016, geologists were able to take core samples from the deeply overlain crater using a drill borrowed from an oil company. The rock proved to be granite, that is, material ejected from deep within the earth, and not the gypsum one would expect to find on an ancient seabed – and so offering further evidence of an impact. In 2019 came more evidence that the event had rapidly acidified the oceans and that the climate remained hostile to life for a long time afterwards. Most scientists are now pretty certain that the great K/T mass extinction was caused by a projectile from outer space. The small brains of the dinosaurs had nothing to do with it (besides which, their intelligence was probably under-rated; some of the smaller two-legged ones might have been almost as bright as an ostrich).

What were the consequences? The dinosaurs became extinct of course, but so did pretty well every animal weighing over 25 kilograms. Even the survivors were much reduced in diversity. Of fifty-nine species of mammal found at the K/T boundary in North America, only four seem to have survived. Of the various groups of ancient birds, only one, the Aves, survived to become the ancestors of all modern birds. Reptiles were reduced to the modern kinds – turtles, snakes, crocodiles, lizards. For years after the impact, the main fossils were fungus spores – representing, one presumes, the zombie mildews that furred the burned vegetation and the rotting corpses (cooked *T. Rex*, recycled by toadstools!). Once the sun broke through those dark sulphurous clouds, ferns rapidly covered the ground – the great 'fern spike' of the earliest Palaeocene. But this was a silent world. It took some millions of years to restore the vitality of ecosystems and create the humid forests and swamps of the early Tertiary, full of small, mostly arboreal life. Yet, only ten million years after the impact, the initial monotony of shrew or possum-like mammals had diversified into the ancestors of bats, rodents, horses, primates, even prototype whales. With the Mesozoic giants dead and gone, mammals conquered the land, sea and air. The Tertiary was their great age. It ended only ten thousand years ago. It is our age now, long or short, whichever it turns out to be.

Would we have survived the mass extinctions? Well, it's a question. Any one of them might have brought about the end of civilisation. Maybe those who survived the convulsions would find some technological solution, basing a new civilisation on the ruins of the earth as our descendants might one day do on the moon or Mars. But it would beg another question: would such a life be worth living? Granted that we are now in the early stages of the Sixth Extinction, will the end of it be anything like the dead sea at the end of the Devonian, or the desert world at the close of the Palaeozoic, or the mildewed ruins at the start of the Tertiary. Who knows? I like Elizabeth Kolbert's answer best: 'You wouldn't want to find out' (Drake 2015).

Vanished giants: the end of the megafauna,
50–10,000 years ago

I once spent a winter holiday in the Everglades of Florida. The sun shone over the wet plains of sawgrass with their islands of palm, live oak and cypress, the waterways sparkled with birdsong, and all was well with the world. For a British naturalist the scale of the Everglades took some getting used to. The marshes went on for miles and miles, the boardwalks and cycle tracks being mere nibbles around the edge of a trackless wilderness. It felt as if nothing had changed there since the dawn of time.

All the same, something was missing. There were plenty of turtles and alligators, and paddling down one of the broad waterways we soon came across a friendly manatee nearly the size of our canoe. It had an unperturbed air about it as if there was all the time in the world for everything. For manatees and alligators, which were among the world's endangered species not so long ago, the Everglades conservation plan has been a triumph. But all the same, the missing ingredient was obvious: where were all the other big animals? In Africa, marshes like this would be home to hippo and elephant, buffalo and reedbuck, cervals and lions, upwards of a hundred species of mammals. In the Everglades the alligators and manatees were the only visible megafauna.

Back-track a million years and this scene would have

looked much more like Africa, and then some. There were mastodons, glyptodonts, giant sloths and other exotica, as well as herds of horses, buffalo, wild cattle and deer. And to hunt them there would have been wolves and big cats, including the lion-sized *Xenosmilus*. But, in common with the rest of North America, and South America too, most of the big mammals disappeared around 10,000 years ago. The habitat remained, but shorn of its megafauna. Perhaps not many generations of humankind ever saw the Everglades as a fully functioning ecosystem with its natural complement of animals. We entered North America as an invasive species, and as *Homo sapiens* multiplied and spread through the pristine land, so the megafauna began to disappear. According to the American palaeo-zoologist Paul Martin, the extinction front followed humankind's trail like an isobar. For the New World it was the start of the Sixth Extinction, the latest mass extinction on earth.

Modern humans emerged in Africa around 200,000 years ago. Around 120,000 years ago they migrated into Asia and Europe. About 70 to 50,000 years later they reached Australia. And just 20,000 years ago, or even later, they crossed the land bridge in what is now the Bering Strait and discovered America. And coincidental with all those dates, life, especially large forms of life, began to disappear. The previous age to our own, the Pleistocene, was a world of giant animals. Some of the elephants were bigger, and were far more diverse, than today, and they inhabited every continent except Australia and Antarctica. There were many more species of horses, bison, camels and rhinos. There were giant forms of deer, beaver, kangaroo, lemur, tortoise, and even a giant swan. Every continent had several kinds of now-extinct big cats as large as lions. There were enormous condors and flightless birds, and animals with no modern counterparts such as ground sloths, glyptodonts and *Sivatherium*, a kind of giraffe with antlers. They are all gone. And we only just missed them. Some had their portraits painted on the rocks and caves by our ancestors: aurochs, mammoths, stately giant deer, forgotten rhinos. You almost feel you could touch them. The last *Sivatherium* might have coincided with the first ploughed fields. Herds of woolly mammoths still survived on remote Arctic islands when they were building the pyramids of Egypt. When Columbus was discovering America, New Zealand still had giant flightless birds as well as a giant eagle to prey on them. We know roughly when all these wonderful species died out. What is endlessly discussed is why – and also *how*. How was it possible to kill off so many species in so short a time?

Being large seems to have been a great idea right up to the point when humans entered the frame. Then, suddenly, it wasn't, and it still isn't. All the same, there's a lot to explain. It seems that hundreds of species died out at roughly the same time, around 15–10,000 years ago, and that they included more than half of the largest land animals on earth. Each species presumably had its own trajectory, a particular set of events that eventually snuffed out its existence. This time there is no evidence of a single titanic upheaval, such as a meteorite, or a major volcanic eruption, that would wipe out dozens of species at once (but read on . . .). So how do we explain the more or less simultaneous extinction of such different animals as elephants and rhinos, big cats and hyenas, fast-running horses and pronghorns, hardy camels, and a long, long list of tortoises, snakes and raptors? They do not seem to have been any less well suited to their environment than their present-day equivalents (after all, after the extinction of wild horses in North America, their domesticated successors, the mustangs, took to the plains with ease once they were reintroduced by Spanish conquistadors). To judge from their remains, many of these lost animals were widespread, successful species. They were doing fine, having survived a succession of ice ages and various natural climatic convulsions and still come out on top. Until, quite suddenly, humankind arrived on the scene and then they were gone.

There is no broad agreement about the death of the megafauna, mainly because there isn't enough evidence. Four broad scenarios have been proposed, which are not, of course, mutually exclusive: climate change (leading to habitat change); some sort of external agency, a strike by a space rock perhaps; one or more disease pandemics unintentionally introduced by man or his accompanying animals and, finally, hunting by humans, the so-called overkill hypothesis. But before going into them, let us look in a bit more detail at exactly what was lost across the world when the Pleistocene, the age of giants, merged into the Anthropocene, the age of man.

The least affected continent was Africa. *Homo sapiens* evolved in Africa, and our hominid ancestors had been roaming the continent for a long while before that: long enough for wild animals to acquire a healthy fear of two-legged hunters. Africa is the only continent today with a seemingly full complement of large animals roaming in landscapes resembling those of the Pleistocene: elephants, rhinos, hippos, buffalo, wildebeest, lions. Nonetheless Africa did suffer extinctions, and many of them. Some 28 lineages of large herbivores died out during the past seven million years, including species of elephant, hippo, giraffe and

antelope, as well as totally extinct groups such as deino-
theres, elephant-like animals with backward pointing tusks,
and chalicotheres, very strange animals resembling a cross
between a horse and a gorilla. Paul Martin estimated that
some 15 per cent of the fauna of Africa died out at the
end of the Pleistocene, a significant proportion though
much less than on other continents. Some very recent
research places the blame not on our human ancestors so
much as on climate change. A global drop in carbon dioxide
levels led to cooling and a shift from dense forest to open
savannah dominated by tropical grasses: exit the species that
could not abide the change. Dramatic changes always result
in extinctions, especially of large animals, but in Africa the
spread of open savannah grasslands also benefitted other
species. In Africa, then, large mammals and humankind
managed to coexist, more or less. For Africa, *Homo sapiens*
has been let off the hook – for now.

In Eurasia there seems to have been a gradual loss of
large animals, starting around 50,000 years ago and culminating around 13–11,000 BP. One determinant was the
loss of an entire habitat: the mammoth steppe, the cool,
grassy plains that once stretched from Western Europe right
across Siberia and on to Alaska. This grassland was maintained and fertilised by enormous herds of large cold-adapted
herbivores: among them woolly mammoth, woolly rhinoceros, saiga and a giant deer, *Megaloceros* ('big horns'),
also known as the Irish elk because many of the first fossils
were dug out of Irish peat bogs. There were also now-
extinct species of horses, musk oxen and antelopes. They
were preyed on by a long-legged, dirk-toothed cat,
Homotherium and a subspecies of lion larger than its present-
day African confreres.

Then, around 12,000 years ago, there was a dramatic change: the open mammoth steppe was replaced by taiga, the endless conifer forests of the far north, unsuitable for mammoths or any other large animal that grazed in herds. The mammoths died out. In Britain, not yet an island, we lost nearly all the larger animals: elephants, a rhino, a wild ass, bison, giant deer, saiga, wolverine, and, on a smaller scale, Arctic fox, lemmings and pika.

The role of the lost grazers was simulated in the fullness of time by domesticated animals: cattle, sheep, pigs and horses. The islands of the Mediterranean lost even more: *all* their larger animals, including dwarf forms of elephants and hippos, giant rabbits and dormice, and that amazing giant swan. All the medium-sized animals that inhabit those islands today – mouflon, wild goat, deer, wild boar – are subsequent introductions. At some point along the line our nearest cousin, the Neanderthals, also vanished, and in circumstances no less mysterious – though we still carry some of their DNA, including, apparently, the gene for red hair.

Humankind first discovered Australia around 70 to 50,000 years ago. It seems we reached the continent without knowing it was there. All that time ago, deep into prehistory, humankind was already an explorer. Australia was once home to giant marsupials, giant birds, and equally giant reptiles in one of the strangest faunas on earth, having evolved during the island continent's long isolation. There was a rhino-sized animal related to the wombat called a diprotodont and its likely predator, the marsupial lion. There was a claw-footed kangaroo taller than a man, a giant koala and a host of weird reptiles: tortoises, pythons, crocodiles, lizards and iguanas. Australia's birds were led by *Genyornis*, much bigger and heavier than an emu. There were also marsupial equivalents of hippos, tapirs, cats and wolves.

From that distant Eden, some sixty species of vertebrate animals died out, including – surely significantly – every species larger than a man. Today's arid continent could never have supported such a diverse fauna, and so habitat change might account for at least some of the losses. But what caused the habitat to change? Some species, such as *Genyornis* seem to have disappeared abruptly, leaving us lots of carbon-datable bits of egg-shell. Perhaps that was climate, but, if so, the same change did not seem to bother the equally giant moas of New Zealand, which survived into the last millennium. So was the present-day aridity of Australia the result of humans, hunting, burning and generally laying waste? When humankind reached New Zealand we quickly finished off the moas, and so perhaps we killed off *Genyornis* too, with similar efficiency, not to mention all those meaty diprotodonts and giant (and relatively slow-moving) kangaroos. Such animals were certainly hunted, but the evidence is too thin for any certainty over how, why, or even when Australia lost its megafauna. All we know is that it did.

Very recent evidence suggests that Australia and New Zealand underwent a potentially lethal period lasting a few hundred years around 42,000 years ago. Oxygen isotopes trapped in fossil trees of that age indicate that the magnetic poles of the earth had flipped, north to south, which temporarily weakened the magnetic field and destroyed the ozone layer. The ensuing electric storms and blasts of ultra-violet solar radiation would have made life extremely unpleasant for a while, especially for large forms of life.

But, if so, we need more evidence from the fossil record to confirm whether or not it resulted in extinctions.

Until recently it was thought that humankind first entered North America at the end of the last Ice Age, about 11,500 years ago. New evidence suggests that people were already there, having arrived several thousand years earlier, although maybe not in great numbers. Presumably humans crossed the then land bridge between Asia and North America in pursuit of game and found a vast new world there for the taking. Within a few thousand years, or perhaps even less, the wildlife of North and then South America was changed forever in the greatest megafaunal extinction of all. Some fifty-one genera of large mammals died out, including all the elephants, those splendid mastodons and mammoths, the last New World camels and horses, all the ground sloths, all the glyptodonts, plus various species of bison, wild cattle and pronghorn. The continent also lost all its big cats except the jaguar and the puma, including the most powerful of them all, the sabre-toothed *Smilodon* and the American lion, plus the biggest bear and the biggest wolf. The survivors were the bears, whose diet does not depend on large animals, plus the last big herbivores in the form of the American bison and musk-ox, the latter avoiding humankind by living in the far north of the continent. Three-quarters of all animals weighing over 100lbs perished. As did a lot of birds including the magnificent teratorns, giant condors that presumably depended on a regular supply of large corpses. It was perhaps the most spectacular fall of large life since the end of the dinosaurs, 67 million years earlier.

It was much the same story in South America. The two halves of the Americas had joined only recently and there followed an interesting intermingling of their very different faunas. North American horses, camels and sabre-toothed cats moved south (the latter perhaps putting paid to the long reign of flightless terror birds in South America), while ground sloths and glyptodonts moved north. But as human-kind moved into the continent in multiple waves, the South American megafauna died out too, including both native elephants and all the ground sloths, such as the fabulous elephantine *Megatherium*, plus lots more strange and wonderful animals like the pampatheres or giant armadillos, the camel-like litopterns and the hippo-like toxodonts. The largest animal in South America today is a tapir the size of a large pig (the llama, which is taller, is a domesticated animal; the wild guanaco is taller but lighter).

Altogether, the American palaeo-zoologist Ross MacPhee estimates the vertebrate losses of the past 50,000 years to be in the area of 750 to 1,000 species – about a fifth of all mammals surviving today. Some of them, like the woolly mammoth, would have been keystone species on which whole habitats and ecosystems depended. Their loss would have altered the natural environment quite radically.

How did that happen?

The pioneers of geology were much drawn to the notion of some great calamity that had destroyed much of life on earth and so accounted for the fossils preserved in the rocks. For believers in the literal word of the Bible, the obvious model was Noah's flood and the animals that he presumably *didn't* take on board the Ark. Later on, the climatic convul-sions of the Ice Age itself seemed like a more plausible explanation. The idea also grew that human beings might have had a hand in Ice Age extinction through over-hunting. But before the discovery of radio-carbon dating, it was impossible to link the loss of the megafauna to any particular event, and so it was all speculation.

As we have seen, previous mass extinctions were caused by sudden radical changes to the world's climate. Did any such planet-wide crash occur at the end of the Pleistocene? There is some evidence of a strike, or strikes, by meteorites, at about the right time, though it is controversial. In 1994, the Austrian geologist Edith Kristan-Tollman proposed that a fireball striking the earth would account for a sudden climate cooling in the thousand-year period starting around 12,900 BP known as the Younger Dryas (named after *Dryas* or mountain avens, a cold-climate plant whose pollen is prominent in sediments of the time). This was 'Tollman's Hypothetical Bolide'. There is evidence of such an event in spikes of platinum within a sediment layer from that time, comparable to the iridium spikes of the K/T mass extinction, and attributable to a meteorite. There are also beads of glass of similar date formed from molten rock. But unfortunately no impact crater of the right age has yet been found.

In 2007, the appropriately named American scientist Richard Firestone proposed that a comet had exploded not on the ground but in the upper atmosphere, scorching the land below with its intense heat rather like an atom bomb. The smoke and dust billowing into the clouds from worldwide fires might, he thought, have been enough to trigger the mini Ice Age and so cause the extinction of vulnerable animals. The fall in global temperature also caused the displacement of forests with tundra, and increased glaciation in the Arctic and on mountain ranges. But there are other, perhaps more plausible explanations for the Younger Dryas, such as a shutdown of the North Atlantic 'conveyor', the system that circulates warm, tropical waters northwards. A northward shift in the jetstream in response to melting ice sheets and the resulting cooling and refreshing of the Atlantic might be enough to account for it – in Britain we know only too well that jetstreams can make or break a summer. Besides, the dates don't fit. Recent research suggests that the megafauna collapsed a thousand years *before* the Younger Dryas, while the woolly mammoth certainly survived during and after it.

A second possibility is disease. The idea that only 'hyper-disease' could account for the loss of so many species at the same time has been championed by Ross MacPhee of the American Museum of Natural History, whose beautifully illustrated *End of the Megafauna* is the most widely read book on the subject. MacPhee believes that hunting by Stone Age tribesmen could not have caused such havoc in so short a time. 'Look at elephants today. They don't stand around waiting to be shot at!' (MacPhee 2019). Disease, as we don't need reminding, can have devastating effects. In the 1890s, rinderpest killed up to 90 per cent of cattle in South and East Africa and was also transmitted to a whole range of wild animals, either by direct contact, contaminated drinking water or from breathing contaminated air. In 2015, a mystery disease killed off two-thirds of the already endangered saiga, one of the survivors of the megafauna in Asia. In Tasmania, facial tumour disease threatens to wipe out the equally endangered Tasmanian devil, the largest surviving marsupial carnivore. The two endemic rats of Christmas Island were quickly wiped out by a disease transmitted by cosmopolitan black rats that had jumped ship: a kind of rat-to-rat Black Death. And without vaccines, Covid-19, a pandemic probably transmitted from bats, would, at the very least, have reduced the life-span of the average human being (as it is, the UK male life expectancy has fallen for the first time in forty years).

MacPhee's idea is that, when modern humans made contact with wild animals for the first time, they would have had little resistance to the diseases that we (or our animals) brought with us. They may even have caught something fatal from our garbage – the rotting remains of previous dinners. Epidemics spread fast and, unlike hunting, can slay tens of thousands in a very short time. It may also explain why, after the loss of the megafauna, no other really large species died out until modern times.

The problem with the disease hypothesis is that it is not susceptible of proof. Perhaps 'pathogenic signals' – evidence for a widespread epidemic – could survive in the preserved bone marrow of Pleistocene animals but identifying specific micro-organisms responsible would be difficult. And, besides, there are other objections. There is no known disease that would affect so many species. Microbiologists doubt whether any bug ever known could be that infectious, or that virulent. The most they will concede is that disease introduced by humans or their agents may well have played a part, especially in the Americas, but whether or not it did and to what extent, we cannot know.

The evidence for climate change is stronger, but once again is difficult to tie down. One of the best researched examples concerns the famous woolly mammoth. Thanks to cave paintings – Ice Age art – we can glimpse what this wonderful beast looked like. I was lucky enough to visit the Altamira caves in northern Spain before they closed. Among the amazingly lifelike beasts I saw drawn on the walls and roofs of the caves was a mammoth, so different from today's elephants in its high shoulders, sloping back, humped head and with enormous, curved tusks. The mammoth was what ecologists call a keystone species, maintaining its habitat by grazing, by its dung and by its heavy tread, which broke the ground and ensured a turn over of fertility. Dozens of other steppe animals would have depended on the mammoth. Today, with no mammoths to maintain them, those vast northern grasslands have mostly gone, replaced by coniferous forest, the endless Siberian taiga, cold, dark and infertile. There's a chicken-and-egg riddle about which came first, the extinction of the mammoth or the spread of the forests? Did the changing climate favour trees over grass? Or did the loss of those great herds propel the spread of trees?

The problem with climate change as an overall explanation is that these animals had all experienced similar changes before – what we call *the* Ice Age was only the most recent one in a long series lasting over seven million years. They had adapted to the cold, for example by migration, had lived through it, and they were still thriving. Tusks of North American mastodons killed by human hunters and examined by modern scientists show no signs of stress. They were healthy animals. They and other big mammals seem to have died out when the climate was no more stressful than before. It is also hard to imagine how climate change would have eliminated, say, the horse. At the most, it would only account for local losses, not a continent-wide mass extinction.

Which leaves humankind: the smoking gun – or, rather, we should say, the quivering spear. The American palaeozoologist Paul S. Martin led the way in gathering evidence to show that wherever human pioneers went, whether in

Australia, the Americas or on ocean islands, native animals died out. In North America, he likened the advance of humankind across the continent to 'a bow-wave of destruction' (Martin 1984), calculating its progress at about 16 kilometres a year. It swept across the forests and plains, into the tropics and down into South America and, at the crest of this human tsunami, large animals dropped from the record. Martin likened it to a blitzkrieg, an advance as lethal and efficient as the advance of the Wehrmacht across France in 1940.

The Australian palaeo-ecologist Tim Flannery noted that the human invaders of North America might have lived in the Stone Age but they wielded very sharp spears – their 'Clovis points' (Clovis, New Mexico, is where these spears were first found). They were tribes from Asia and the ancestors of modern Native Americans. The Clovis people left very little figurative art – there is no equivalent rock art of North American mammoths or sabre-toothed cats – but they were evidently expert hunters. 'Theirs', Flannery feels, 'was an aesthetic that would find beauty in a well-oiled engine' (Flannery 2001). He likens the fall of the megafauna of an entire continent to light falling into a black hole, a hole in this case lying between the Clovis nose and the Clovis chin: a big, greedy open mouth. In other words, Clovis and his chums destroyed the megafauna by hunting it and then eating it.

One reason why these animals could be hunted so successfully was that they probably had no fear. Tameness in a wild animal, MacPhee has noted, 'is a death sentence' (MacPhee 2019). They were given no time to acquire a terror of their hunters, for instinctive fear takes centuries to evolve. Part of the pleasure of birdwatching in New

Zealand is that, even after several hundred years, many of the indigenous birds don't fly or run away on contact. They stay and investigate your pack and sometimes walk away with your lunch. By the same token one can imagine these now-extinct American animals staring stupidly at the circle of Clovis points, or perhaps approaching a campfire full of innocent curiosity. And one doesn't even need to imagine the fate of fearless island animals and birds, like the dodo or the last ground sloths on Caribbean islands, killed off within a generation or two of the arrival of the first ships.

All the same, there are objections to this thesis. To begin with, where are the bodies? Most of the kill sites excavated in North America are of the bones of mammoth and mastodon, perhaps of animals lured into boggy ground, trapped in pits or even driven over a cliff. The Clovis people hunted bison, just like their descendants on the plains, but there is little evidence that horses and camels were widely hunted, while the ground sloths might have been quite dangerous adversaries, with their thick skins and powerful arms and claws. Wiping them all out in so short a time (which might explain the lack of kill sites) would require extraordinary levels of slaughter. Were those horse-less tribesmen really so numerous, so skilful – and, above all, so greedy? Perhaps, but the mind still boggles at *how* they achieved this – after all, extermination wasn't their *aim*. Those big beasts were useful. A recent study (Smith 2018) argues that you don't actually need to go all out to wipe out a species. All you need is a population that is already stressed and a level of hunting that is just sufficient to lower the fertility rate below replacement levels. In those circumstances population collapse becomes mathematically certain.

We should also factor in environmental damage caused by humans, such as fires, disease and habitat stress, too. Although the extinction of the North American megafauna was unique in its scale, humankind has proved well capable of exterminating large animals in historic times, such as the leopard or the lion in Europe, or rhinos over most of their range in Europe, Asia and North Africa, well before the introduction of firearms. Once an animal's population has been squeezed into isolated pockets, weaknesses from inbreeding and environmental stress make them more vulnerable to overkill. All the same, all the theories that attempt to explain the death of the megafauna are based on conjecture. What actually happened remains a mystery. You could imagine it as a Who-dunnit game, a sort of Megafaunal Cluedo. It might be him, it might be her; it all turns on which evidence cards you hold in your hand.

The reason why some big North American animals *did* survive, including at least two that were most definitely hunted by the Clovis tribes, bison and elk, could be due to the fact they were immigrants. Bison, elk or wapiti and the grizzly bear all came from Asia and already knew that *Homo sapiens* was a beast best avoided. The grizzly may also have been able to move into the space lately evacuated by the extinct short-faced bear, just as the grey wolf may have exploited the niche left by the extinct dire wolf.

So was it really us?

'*Homo sapiens* appears unique because of its [sic] ability to exterminate other species', writes Jordi Agusti, author of *Mammoths, Sabertooths and Hominids* (2002).

It seems that the process of extermination started early,

indeed, almost at once, once we'd left the cradle of hominid evolution in Africa. Our intolerance of big beasts probably extended to our fellow hominids, the Neanderthals in Eurasia, the Denisovans in Siberia, the tiny 'hobbit' man on Flores and maybe also the archaic hominids in Africa some time before that. All our fellow hominids are long gone. Now, the survival of the next group down the line, the great apes, is threatened: gorillas, chimps, orang-utans, bonobos, every one of them.

The trouble is that while investigation can sometimes detect the approximate date of extinctions, it won't tell us the whole story. Extinction is not always a relatively simple, linear process, a straight slide into oblivion, but a more complicated series of curves, a continual adaptation to change in which a species might decline, but then perhaps rebound, but each time moving just that bit closer to the edge. If we could be carried back to the late Pleistocene in a time machine, we might see the great beasts apparently prospering, just as rhinos and elephants in Africa were until very recently. We would probably not be able to tell the winners from the losers. We might also ask ourselves which of those Ice Age animals would survive today. Sabre-toothed cats? Probably not – there isn't enough big, slow-moving prey to feed them. Woolly mammoths? Again, debatable – their habitat has vanished. Extinct wild horses and camels? Giant kangaroos? Sure, why not? If we gave them a state to live in, say, the whole of Wyoming or a recently dehumanised New South Wales.

Humankind is implicated in the death of the megafauna through the sheer weight of coincidences: the arrival of man seems to be followed by the collapse of the megafauna throughout. That evidence appears most certain on islands, such as New Zealand, Mauritius and Madagascar, but is also fairly compelling in the Americas. However it happened, it has left the world more denuded of large land animals than at any time in the earth's history. Yet, without the evidence in the rocks or buried in the earth, it would be easy to assume that the wildlife we see today is just the natural order of things: five types of rhinos, two or three elephants, two camels, one hippo, five big cats. Big animals, it would seem, are rare. But in the natural, pre-human state, the megafauna wasn't rare: it was everywhere, in herds (or stalking the herds), in the swamps, the open forest, the grassy plains, even along the shore. In truth, we have almost succeeded in replacing big animals with one animal: the human being, on the streets, in the parks, along the roads and byways, on the beach. The age of giants is over. We live in the age of us.

Chapter 2

Going, going ...

Biological diversity is messy. It walks, it crawls, it swims, it swoops, it buzzes. But extinction is silent, and it has no voice other than our own.
    Paul Hawken (1993), *The Ecology of Commerce.*

So much for the past. What about the present?

While 99 per cent of life that existed may be extinct, there remain millions of species that share the world with us today. How many millions exactly? No one knows. We can obviously only count the species that have been discovered and scientifically described (and for which there is physical evidence in collections). That number is around 1.9 million, yet it is nowhere near the likely real figure.

To begin with, the majority of species are not mammals or birds but far less prominent life forms: invertebrates, algae, fungi, bacteria, and we are still only tapping the surface of *that* vast biodiversity. There is also a considerable fluidity of opinion as to what actually constitutes a species, an argument that deepens as you descend from the tiny into the microscopic. New species are being discovered every day. Recently some 16,000–18,000 organisms new to science have been described each year, most of them microscopic. The world's actual theoretical biodiversity is anybody's guess but it is believed to lie somewhere between 15 and 30 million species. In other words, most living things on our planet haven't been discovered yet. Many of them never will. At the present rate, even the lower figure of 15 million would take us a thousand years to complete. And, of course, we will never complete, because a great many species will die out before the scientists manage to find them.

If we stick only to the better-known species, that is the vertebrates and a selection of plants, insects and other invertebrates, we are in more manageable territory. But even here we do not know the exact number because new ones are being discovered all the time. There are, for instance, 43 per cent more amphibian species in the world today than there were 20 years ago, thanks to a combination of discovery and refined species concepts. An average of 150 amphibians – mostly frogs – are being discovered every year. Even the world's mammal list has increased by 20 per cent during the past decade or so. The exact number of species for any group vary from year to year and from source to source. The latest figures collated by the World Conservation Union are 5,513 mammals, 10,425 birds, 10,038 reptiles, 7,302 amphibians and 32,900 fish – and about a million insects. These figures will be out of date by the time you read this.

Meanwhile species are disappearing all the time. The figures for recent extinctions – 'recent' being loosely defined but let's say from 1800 onwards – are uncertain. The main difficulty lies in proving that a species really is extinct. The official figures, such as 540 lost vertebrates, 96 lost mammals or 158 defunct birds, are only the ones we know about. They are almost certainly a vast underestimate. For example, it is reasonable to suppose that most of the isolated islands of the Pacific were once home to at least one endemic species of bird. There are hundreds of such islands and that means hundreds of lost island birds. Most of them will have died out without anyone noticing because, after all, the island settlers weren't ornithologists. Many species currently listed as 'critically endangered' haven't been seen by anybody for many years. They may well be extinct, the thin end of the wedge of the Sixth Extinction, but it is too early to be absolutely sure.

As for the world's plants, it is impossible to even guess at how many have been lost. According to the International Union for the Conservation of Nature (IUCN), 108 species are 'probably extinct' and 37 are 'extinct in the wild' (that is, lost from nature but retained in cultivation). A further 872 species are considered to be critically endangered, and a startling 3,654 more species are on the danger list. On top of that, some 1,674 more plant species are 'Data Deficient', which means they are probably endangered too, but no details are available.

The number of species in the world that are believed to be *threatened* with extinction is enough to make anyone blench. For example, of the 7,000-odd species of frogs and toads, it is thought that we may well lose nearly half by the end of the century. Some 1,244 species of mammals are believed to be threatened, in 200 cases critically so, and they include some of the best-known and most beloved animals on earth: rhinos, the great apes, gibbons, lemurs, marsupials, big cats. For birds, the best recorded animals on earth, some 1,480 species are threatened, critically so in 223 cases, and, again, that includes some of the world's loveliest species: parrots, pigeons, rails, hawks, owls, hummingbirds, woodpeckers.

'Critically endangered' means that unless some worth-while protective measures are taken pretty fast these species may well soon be gone forever. The label 'vulnerable' means that numbers are falling at a worrying rate but that extinction is not yet inevitable. But, for the most part, these are only guesses. For some species, especially those in wealthier parts of the world, considerable measures, including captive breeding and translocations, are being made to save them. But such cosseting is almost confined to attractive and popular species. We care about tigers and ospreys and spend a great deal of money on them. We don't care, or at least not so much, about bryozoans or barnacles or seaweeds.

Wondering how to tackle such huge numbers – for to list the world's threatened vertebrates alone would take up half this book – I have chosen to focus on four examples to illustrate what is happening to the wildlife of the world. First, I have chosen two large but contrasting islands where the extinction rate has been extremely high. Second, I review a selection of the world's most endangered animals and birds to see how they have fared during the past sixty years; this is a tale of hope as much as heartbreak. Third, I review the plight of those much less sung but far more numerous entities, the world's insects. And finally, I focus on just one family of highly endangered animals: the rhinoceroses.

Two islands

Lost Eden: New Zealand

I once spent half a winter in New Zealand and managed to see many of the wonderful birds that live there. What remains of New Zealand's natural vegetation is perhaps the closest we are ever going to get to the real Jurassic Park, to a landscape of ancient trees isolated from the rest of the world for the past 85 million years. Before humankind discovered New Zealand it had no mammals of any kind bar a couple of bats (one of which, weirdly, hunts by scurrying along the forest floor). Instead of mammals, it had birds, many of them flightless, and some very large. The signature bird of New Zealand today is the kiwi but, had they survived, we would wonder more at birds hundreds of times larger, at the giant moas, some of which were taller than an ostrich.

The moas were huge and stately, but their existence depended on isolation from the other world of large, fierce mammals – and humankind. They strode through forests of tree fern, pohutukawa, podocarp, cabbage tree and the towering kauri, called by the Maori *Tane Mahuta*, the God of the Forest. The moas were New Zealand's megafauna. Some plants evolved large berries and seeds that could be swallowed and dispersed only by moas. A whole ecosystem depended on them.

The endemic birds of New Zealand are the most endearing I have ever encountered. The local robins peck at your shoelaces. Disembarking from the boat on the island of Tiritiri Matangi, a short boat ride from Auckland, we were met and followed by a takahe, a kind of giant, flightless moorhen, a feathered harmony of silken blues and greens, with a huge and intimidating red bill. It acted as if it was showing us *its* island. On another island an equally flightless weka, an elegant, chicken-sized land rail, ran off with my lunch bag. Some birds have songs that make you stop and stare. When I met the kokako, the blue-wattled crow, the actual bird was hidden in the vegetation, a dense scaffolding of fronds, but what I heard erupting from the tangle sounded like a bird concert, a series of creaks, pops and yelps, followed by a most unexpected wheezing noise that filled the air. It sounded as if the kokako was playing a mouth organ!

The kokako, like several other endemic birds of New Zealand, was saved from extinction at the last minute. When Europeans joined the Maoris in settling this distant land, so like, and yet also unlike, Britain, they stocked it with their favourite animals: sheep, cattle, goats, pigs, cats, hedgehogs, rabbits, sparrows, magpies, possums, ferrets and, less deliberately, rats. It helped them feel at home; it should have been New Britain, not New Zealand, after all. For birds, whose entire existence depended on freedom from greedy mammalian predators, the effect was immediate and catastrophic. Grasslands for sheep replaced their forest habitat, the rats and hedgehogs ate their eggs, the stoats their nestlings and cats caught and ate the adult birds. One lighthouse keeper's cat caused an extinction all by itself; it caught and ate all the Lyall's wrens in their last refuge on earth. The birds were flightless and the cat was named Tibbles.

Since there was no way these birds could live alongside voracious mammals, the only alternative to a life in captivity was to find places where animals, especially rats, could be excluded. Fortunately, New Zealand has many small offshore islands which the rats and other pests were either unable to reach or could easily be trapped and the area secured. Tiritiri Matangi, newly planted with native vegetation, where I first encountered the takahe, was one such island. My encounter with the takahe was quickly followed by ones with a host of other bizarre birds, found nowhere else in the world: the saddleback, stitchbird, bellbird, kokako. Had I been able to spend the night there I might even have heard the call of the little spotted kiwi, a ball of mottled fur-like feathers emitting a sharp, curlew-like trill close to the conservation volunteers' hut. Visits to Tiritiri Matangi, Kapiti or one of the other safeguarded islands are a privilege, a stroll through an earthly paradise, albeit human-made, created through hard work, trapping, planting, translocating and stocking.

For some birds, conservation came too late. By the time the first European ships were anchoring in New Zealand's beautiful bays, in the sixteenth century, the great moas were already gone (along with the giant eagle that preyed on them) and after the extinction of the moas, it seems possible that the eagle resorted to attacking the only other large, two-legged animal available – man. Moas had never before encountered human beings, nor their dogs. One imagines they were tame and possibly unaggressive and so, easily despatched. Such an easy source of meat fed a growing Maori population which in turn required more and more land under cultivation, and so diminished the moas' natural habitat still further, giving them no space to hide. With the moa perished, other birds were unable to adapt to the new, far more stressful circumstances. They included a flightless pelican, a flightless goose and an extraordinary crane-like bird with a downward-pointing beak called an adzebill.

Flightless island birds are always vulnerable and among the first to die out. The arrival of Europeans in New Zealand, along with their cache of animals for 'naturalisation', further impacted on the native birds and propelled them even nearer to extinction.

As the native bush was cleared, slashed and burned into fragments, another round of birds disappeared: New Zealand quail, Auckland Island merganser, a thrush called a piopio (from its song), and the Stephen's Island wren. Perhaps the saddest loss of all was the whekau or laughing owl. Though it was more often heard than seen, according to local lore, you could attract its attention by playing the accordion. The laughing owl fed on the Polynesian rat, the kiore, now also extinct. It was best known by its call – described as a weird, maniacal cry, like two men crying coo-ee to one another from a distance. That was the last sound you heard of the whekau: its laughter.

The bird I would most love to resurrect is the huia, a relative of the kokako, which was last seen inside its dwindling patch of pristine forest in 1907, though it may have survived for a little while longer. The huia was a beautiful bird with black feathers, augmented by bright orange wattles and a white-banded tail, which shone in the sunlight with a greenish iridescence. While the male bore a crow-like beak, that of the larger female was curved downwards like a curlew's, enabling her to probe rotten wood for beetle grubs, a valuable source of protein needed for her eggs. This 'bill dimorphism', the greatest of any bird in the world, gave the huia its humorous generic name, *Heterolochia*, 'different wife'.

The huia enjoyed an honoured place in Maori culture and traditions. Unfortunately that honour took the form of killing it for its feathers, while the skins were used by the Maori to make hats and scarves. Perhaps the huia would not have lasted much longer, in any case, for its old forest habitat was disappearing. Like other famous birds the huia has an afterlife, appearing on stamps and coins. There is a huia winery, a huia publishing house, and a huia suburb in Auckland. A single feather auctioned in Auckland in 2010 fetched 8,000 New Zealand dollars (about £4,220). Unfortunately, we will never resurrect the huia, nor learn anything more about its habits. Any DNA that might have been preserved in its bones and fading skins has perished.

New Zealand's lost birds give a powerful echo of a former Eden, a place of evolutionary wonder that we missed by only a few hundred years. How long, one wonders, did the moa survive its first contact with human beings? A few hundred years? Or was it just decades? The second round of extinctions, during the Western settlement of the nineteenth and early twentieth centuries, including the huia and the poor laughing owl, might have been prevented had conservation entered political discourse a few decades earlier than it did. Without it, New Zealand would probably have lost a third round of native birds, leaving the place to the chaffinches, sparrows and blackbirds imported halfway around the world by the mother country of the Commonwealth.

One of the species which the European settlers brought with them was *Bombus subterraneus*, the short-haired bumblebee, which it was hoped would cheerfully pollinate the crops of apples, beans and fodder clover. Whether or not it did, the bee thrived in its adopted country but – oh the irony! – it subsequently died out in Britain. Hence, in a reverse journey halfway around the world, New Zealand bees were chosen to restock their former home. One hopes they travelled by way of America thus completing their generational circumnavigation. Unfortunately, the experts forgot that New Zealand summers are wintertime in Britain, and the experiment failed. But, thanks to the least harmful of those nineteenth-century 'naturalisations', a lost bee that was common enough in Queen Victoria's England continues to buzz contentedly among the Hebe and Manuka trees of an island twelve thousand miles away.

Bird conservation in New Zealand today is a highly developed art. The last-ditch solution of removing birds to offshore islands has been successful in preventing another round of extinctions, but the populations are small and the removed species are still considered to be endangered. Fortunately for them, New Zealand is a relatively wealthy country with a low and stable population (5 million, about the same as that of Yorkshire). Concern over the loss of its special birds is shared by the people and government alike. That is why the country has embarked on a very ambitious scheme to rid the mainland of rats, stoats and possums by 2050: the 'Predator-Free 2050' project (ironically the same possums that are such pests in New Zealand are endangered in their native homeland of Australia). Declaring war on rats will cost an estimated nine billion New Zealand dollars, in annual tranches, administered by a linked company. It will generate novel ways of trapping and killing the unwanted animals, including the possibility of using 'gene-drives' to create genetically modified rats ('Imagine giving all the rats in New Zealand a peanut butter allergy, and then we feed them all peanut butter', dreamed one advocate. Yong 2017). Gene-drives are controversial, as, to some extent, is the intention behind the project – is this 'ecological xenophobia', war on the settlers, on the strong and successful? If it succeeds, and it must be a big 'if', then the island's surviving resident birds will have a much rosier future. Had it survived, I would even have put money on the revived fortunes of the huia, if not the moa.

Madagascar: the world in a crucible

I have never been to Madagascar, the other island large enough to support its own unique megafauna, but I know conservation-minded people that have. They returned charmed by the island and its friendly people, inspired by its beautiful scenery and wildlife, and deeply concerned about its future.

The world's fourth largest island, larger than France and more than twice the size of Britain, Madagascar has been isolated for even longer than New Zealand, surrounded by deep sea since the Cretaceous, 88 million years ago. Madagascar separated from the continental mainland at a primitive stage in mammalian evolution and has long been the last earthly home to the most ancient of primates, the lemurs. Like New Zealand, the island was also home to giant flightless birds, in this case the elephant bird, *Aepyornis*, which weighed up to 730 kilograms and laid the world's largest eggs, around 18 pounds each and containing as much yolk as 160 chicken eggs. There were also native crocodiles, giant tortoises, dwarf hippos and the world's largest collection of chameleons. Some of Madagascar's invertebrates are equally bizarre; big day-flying moths as bright as butterflies, giant hissing cockroaches and a spider that makes the world's toughest webs. Including invertebrates, Madagascar may be home to 100,000 species found nowhere else on earth.

Being close to the mainland of Africa and also on trade routes from the Cape to India, Madagascar was discovered and colonised much earlier than New Zealand. The island seems to have been settled in two main waves: from Indonesia, about 2,000 years ago, and from the African mainland, about 500 years later. Human settlement led to a gradual conversion of the island's natural vegetation from forest to open grassland dotted with trees, including the baobabs, signature trees of Madagascar, with their smooth swollen trunks and tufted crowns (when leafless they look as if they are growing upside down with their roots in the air).

There are no large animals left in Madagascar: the island has no native animal larger than a fox. So what happened to all the elephant birds, the giant tortoises and crocodiles? Where are the hippos, the bigger lemurs and their predator, the giant fossa?

Extinctions probably started soon after the island was settled. Climate change, which is implicated in the loss of the Australian and American megafauna, does not seem to have been a factor here. Madagascar, two thousand years ago, seems to have been a lush, balmy place. That its largest animals were hunted down for meat is clear from the many cut bones that have been found there. Habitat destruction and the introduction of non-native animals, including dogs, also played their part in what happened.

Since Madagascar is a large island, in many cases extinction seems to have been a long drawn-out process. The largest lemurs went first, in rough order of size. At least seventeen species are now extinct (though it might be as many as forty, depending on how one classifies the remains), compared with around fifty species surviving today. The largest and most endangered living lemur is the indri, which is the size – and frankly has the appearance – of a big teddy bear. Fully grown, it weighs in at about 6.5 kilograms.

In contrast, two thousand years ago, the biggest lemur, *Archaeoindris*, weighed a hundred times as much. It was about the size of a gorilla. Madagascar's lemurs underwent a remarkable evolutionary radiation. There were lemur equivalents of baboons, the aforesaid gorilla-lemur, another big species, *Megaladapis* which resembled a giant koala and even a whole sub-tribe of lemurs, with long arms and legs, which seem to have hung from branches like sloths. At the other end of the scale there were, and still are, lemurs no larger than mice. Once they had to share the island with *Homo sapiens*, lemur size became a liability. Large, conspicuous, perhaps slow-moving, lemurs were walking larders of meat. They had other problems too: they were most probably active by day, had a slow reproductive rate and had insufficient time to acquire a healthy fear of humankind.

The first to go was the biggest, *Archaeoindris* the gorilla, followed by *Hadropithecus*, a ground-living, baboon-like species, *Babakotia*, one of the medium-sized sloth lemurs, and *Megaladapis*, the koala-lemur. *Palaeopithecus*, a slightly smaller sloth lemur survived into early modern times, perhaps late enough to inspire the legend of the *tretretretre*, half-man, half-monkey, as large as a two-year-old calf, with a human-like face. With them vanished the island's only big carnivore, the wolf-sized giant fossa, and a whole tribe of birds: a crowned eagle, a shelduck and a shelgoose, a lapwing, a rail and a cuckoo. The last elephant birds perished around a thousand years ago, leaving behind fragments of eggshell tough enough to use as bowls. The dwarf hippos were thought to have disappeared at about the same time, but some bones were recently carbon-dated to only 200 years ago, indicating that a small number had somehow managed to survive without drawing attention to themselves. They too seem to have inspired legends of a mysterious animal called *omby-rano* or *laloumena*, a dark, plump, grunting animal the size of a cow which, when threatened, fled to the water and swam away.

Still surviving are numerous medium-sized and small lemurs, plus thirty equally unique species of tenrecs or otter shrews. The island still claims a fashion show of colourful reptiles, 290 species of frogs and other amphibians, plus unknown numbers of endemic moths, crayfish, crabs and snails. All these species, from mammals to snails, are confined to Madagascar, and so their extinction would be global and for keeps. What are their chances? Lemurs and some other species are legally protected. They appear on lists in offices, and annexes to treaties and laws. However in rural Madagascar protection may not mean very much because the people probably don't know they are protected. Local taboos may be more meaningful than international agreements because they tie in with the indigenous culture. Madagascar also has many national parks, and it has quadrupled the area of legally protected land since 2003. Some parks are visited by tourists and bring in much-needed income (though a drawback is the lack of a good nationwide road network; there were no paved roads at all on the island before 1895, only tracks, and even now it is mainly dirt roads that turn into storm gullies and mud-baths every time it rains).

All the same, in 2020, Madagascar's environment minister, Dr Vahinala Raharinirina, went on record that she thought the strategy of protected areas had been a failure. Her point was that that the parks had not improved the lives of local people, that they were not seen by them as an asset, and so future approaches should be far more 'people-centric' than at present (News Mongabay 2020). Moreover legal protection has not prevented 'alarming levels' of illegal logging and habitat destruction even in the national parks, and nor has it prevented the ongoing decline of lemurs and other wild animals. The IUCN now considers Madagascar's lemurs to be among the most endangered animals in the world.

The overall driver is population. Madagascar is a poor country – 70 per cent of the people live in poverty – with a rapidly expanding population. In 2020, the population was 27.7 million, but increasing at roughly a million babies per annum. On average, Malagasy women have four children each. The present population is expected to nearly double to 54 million by 2050, and quadruple to 100 million by 2100. The demands this will place on the island's economy and resources will be enormous. In biological terms, it will far exceed the likely carrying capacity of the island, and so be unsustainable. Even after the total destruction of the remaining natural forest, the island could not possibly feed so many. Add to that drought, political instability, high levels of debt and the natural wish of people to better their lives, and it is hard to foresee a bright future for the lemurs, tenrecs and chameleons. Some lemurs do quite well in captivity, and the attractive ring-tailed lemurs, in particular, have been widely captive-bred and even sold as (rather problematical) pets. But other lemurs, such as indri and sifaka, do not survive in captivity and, besides, what is captivity compared to an animal's natural habitat? It will take a lot of hope, a lot of money, and quite a bit of imagination to ensure a long-term future for Madagascar's unique wildlife.

Fifty species

All the same it is only human to seek for signs of hope, for reasons to be cheerful. Perhaps I can offer one such sign from a set of fifty picture cards issued in 1963 by Brooke Bond tea on behalf of the then World Wildlife Fund (WWF, now the Worldwide Fund for Nature). It was titled *Wildlife in Danger*. The text and pictures were by Peter Scott, one of the founders of the WWF. Of the fifty chosen species, twenty-three are mammals, nineteen birds, four reptiles, two amphibians and two insects. Some of them are well known – the blue whale, the white rhino, the tuatara, the leatherback turtle. Others, like Clarke's gazelle, the Laysan teal or the giant wood-boring beetle of Fiji, less so. Personally, I have seen only eight of them in the wild, plus a few more in zoos and collections (and I've eaten only one of them – a delicious fillet of farmed bontebok in Cape Town). Back in 1963, Scott and the WWF considered all these animals to be threatened and, in many cases, they were pessimistic about their chances of survival except in zoos and aviaries. So, how are they all doing now, almost sixty years on?

Surprisingly well, in many cases. On my computation, twenty-eight species are doing better now than they were back then, eight are doing worse and twelve are about the same. Just two, the kouprey, or Cambodian forest ox, and the North American ivory-billed woodpecker, have died out. The last universally accepted sighting of the woodpecker in the United States was in 1944. That said, it may still exist, and habitat restoration projects are underway, just in case. A similar bird in Cuba, which may or may not have been the same species, was last seen in 1987. The kouprey, always a very elusive animal, was last seen in the wild in 1970, and there are none in captivity. Both species are still listed by the IUCN as critically endangered in the now remote chance that a few individuals still survive.

Given that the world has less than half the natural habitat that it had back in 1963, a recovery rate of at least 60 per cent for this sample is not bad going. For many species, recovery has been assisted by captive breeding. In the case of the Californian condor (card no. 30), the entire remaining population of just 27 individuals was rounded up in 1987. Today there are 518 birds either in captivity or reintroduced into the wild. It is not yet out of danger but the chances of recovery seem promising. The same is true of the ne-ne or Hawaiian goose, saved by Peter Scott personally at his wildfowl collection at Slimbridge and reintroduced to Hawaii in 2004. Similar captive breeding and reintroductions have been the salvation of bridled nail-tailed wallaby (card no. 2), milu or Pere David's deer (card no. 17), Arabian oryx (card no. 21), Japanese crested ibis (card no. 27), Swinhoe's pheasant (card no. 31) and whooping crane (card no. 32).

This apparent success must be qualified. Few of our fifty species are out of danger, and in too many cases their situation remains parlous, with low numbers and a dwindling habitat. For example, although Leadbeater's possum (card no. 3) is on my 'saved' list, because its numbers did increase between 1961, when the species was rediscovered, and the 1980s, its situation remains critical. The possum has been closely studied in its limited range of mountain forest, generating dozens of peer-reviewed scientific papers, but its habitat continues to shrink from a combination of logging and bush fires, and its numbers are now falling again. A study in 2014 gloomily predicted that the forest ecosystem in which the little tree-dwelling possum lives has a '92 per cent chance of collapse' within the next fifty years. Not good then. It is listed now, as then, as critically endangered.

So, too, is the kakapo or owl parrot (card no. 37) of New Zealand, but its story has taken a different path. As one of the most charismatic birds on earth, our only surviving flightless parrot, and also the world's largest, heaviest and longest-lived parrot, its extinction would be a tragedy. Once found all over the North and South islands of New Zealand, it suffered the familiar fate of persecution – the colonisers had great fun hunting kakapo with dogs – and habitat destruction. By 1963, there were fewer – perhaps a lot fewer – than a hundred individual birds left. A comprehensive survey, in 1995, found only half that many, all of which were rounded up for a serious attempt to save the species. They were relocated to various offshore islands where predator control was possible. But even so recovery was slow, for the kakapo lays a single egg only once every two to five years (chiming with mast years for the fruits of the rimu tree – they are that picky).

Nothing was left to chance. Every bird was fitted with a mini radio transmitter so that its journeys within its new island home could be followed. Every bird is given an annual health check. Their diet is supplemented with carefully weighed amounts of commercial parrot food. Every nest is surrounded by 'poison stations' in case of rats, and the health of each chick is monitored regularly, like a patient in a ward. They are probably the most pampered wild birds on earth, although 'wild' is surely a relative term here. And the consequence is that the kakapo has prospered, but again only relatively speaking. The adult population now stands at 209. The year 2020 saw the best breeding season to date with 200 eggs laid and 72 chicks fledged. On that basis it was hailed as New Zealand's Bird of the Year. But as if to underline the fickleness of fate, in 2019 the kakapo was struck by an epidemic of aspergillosis, a lung disease. 36 sick parrots were flown by helicopter for treatment to one veterinary hospital, 10 to another. The Kakapo Recovery Programme, running on well-greased financial wheels since 1995, cannot afford to lose them.

At what point does a wild species become, in effect, domesticated? Perhaps when the survivors are given pet names? The celebrity among kakapos is Sirocco, a randy 23-year-old male which famously once attempted to mate with a human head belonging to Mark Carwardine, an act described by his companion, Stephen Fry, as 'the funniest thing I have ever seen' (painful though: those sharp claws, plus a beating from its stubby little wings). Sirocco has since toured the country, thrilling his many fans with his live public appearances, and been given a special seat in his plane so that he can peer out and watch the world go by. Sirocco, we are told (and it is easy to believe), is an ardent supporter of kakapo conservation. But, at the same time, he is useless for breeding purposes because, being hand-reared, he is imprinted on human beings, not on fellow kakapos (which is of course the reason he found Carwardine so attractive). His job now is not to continue the species but as a kind of species ambassador, as a world-famous celebrity parrot.

A shared characteristic of Peter Scott's endangered menagerie is attractiveness. For a set of cards aimed at educating children, he wasn't going to pick boring animals, was he? Nearly all of his choices are big, beautiful or elegant and of particular interest to the WWF. We are naturally drawn to conserving wildlife attractive to us, and big, beautiful animals receive the lion's share of funding (in Britain, ospreys, kites, sea eagles and beavers count for more than all the insects put together). Millions have been spent on rounding up, rearing and even tutoring the likes of kakapo, Californian condor and whooping crane. Zoos of the world compete to breed tapirs, lemurs, tuatara, pigmy hippo. The Cape mountain zebra (card no. 10) has its own National Park. And some species also have the power of symbols. The cahow of Bermuda (card no. 25) might look like just another petrel, but, as a 'Lazarus species' once thought to be extinct, it represents a beacon of hope for nature conservation worldwide (see p.224). The cultural importance of the kagu of New Caledonia, a beautiful blue-grey and near-flightless bird (card no. 34), or the Manchurian crane (card no. 32), an age-old symbol of long life, allows them to jump the queue for always limited resources. So the relative success of so many of Peter Scott's fifty may offer a misleadingly rosy picture of conservation. They were selected to impress, and to create concern that animals and birds as magnificent as these may soon vanish from our lives. We can be grateful that most of them haven't yet.

For a still more optimistic assessment, we could turn to another set of fifty picture cards released in Canada in 1970 and titled *North American Wildlife in Danger*. With text by Roger Tory Peterson, America's best-known birdwatcher, it draws on a 1960s' view of the likely future, when the use of pesticides such as DDT was a major concern. The set consists entirely of mammals and birds apart from a lonely butterfly. How are *they* doing now? Are there benefits from living in one of the world's richest countries? The answers seem to be yes and mostly doing quite well, thank you. Most of the fifty are in better shape now than they were in the 1960s. Some – pelicans, osprey, peregrine, alligator, sea otter, wild turkey, bald eagle – are out of danger, and are now officially listed as 'least concern' or even 'recovered'. In the twentieth century North America lost only a single species of mammal (the Caribbean monk seal, see p.175). On the other hand, this second fifty, like the first, are all charismatic animals and birds of high national prestige. A recent, more sober, assessment put the number of US plants, animals and insects at risk at 1,278 species because the problems associated with their well-being continue to grow.

One reason why so many American animals and birds have bounced back since the 1960s was the eventual ban on DDT and its derivatives – thanks to Rachel Carson's alarm call in *Silent Spring*. Another is that there are more resources, both scientific and financial, for conservation in North America than anywhere else on earth. Of course it also helps to have plenty of space. North America is huge; it has national parks bigger than whole English counties. Impoverished countries with fast-rising human populations, or on small islands that may soon be under water, will have a different story to tell. And yet it is in these places where much of the world's wildlife happens to live.

A million invertebrates

Invertebrates die quietly. We rarely notice the disappearance of, as American biologist E. O. Wilson put it, 'the little things that run the world' (Wilson 1987), unless they are being monitored for some reason, or closely studied by someone. But, of course, invertebrates will die out when they run out of living space, just as vertebrates will. In their case the official figures are such a vast under-estimate as to be almost meaningless. Officially (the statistics are from the IUCN in 2016), 395 species of invertebrate have become extinct in the recent past, plus another 206 which are 'probably extinct', meaning that they have been looked for but not found. A further sixteen are 'extinct in the wild'. They include examples from nearly every group of invertebrates on land or in the water. Among the official lost invertebrates of the world are thirteen beetles, nine spiders (plus another sixteen that are 'probably extinct'), four caddis-flies, three millipedes, three earthworms, three grasshoppers, just one dragonfly and no centipedes at all. The most surprising aspect of the IUCN's list is that the majority (311) of recorded invertebrate extinctions are not insects at all but molluscs, especially single-shelled land snails or gastropods. There are reasons for this odd disparity which we will come to in a moment. The majority of known small-life extinctions are from small islands, especially in the Pacific. It is easier to be sure of a loss on a small island than in the vastnesses of a continent.

According to a widely read paper by Jeremy Thomas and associates (Thomas 2016), the most extinction-prone invertebrates ought to be butterflies. In Britain, butterflies have disappeared from 13 per cent of the spaces they occupied in the first butterfly atlas, published in 1984. Thomas suggested that the world losses of butterflies must be equally significant – so much so, that he felt able to assert, when the idea was still a novelty, that the Sixth Extinction had already started and that butterflies were among its early victims.

On the other hand, the number of *known* extinctions of butterflies is still very small. As far as our knowledge goes (which may not be very far), only 5 out of about 17,500 species have gone for good, and only one of those is more than a footnote in butterfly books. That exception is the Xerces blue, a pretty little American butterfly which had the ill luck to live on the Californian coast near San Francisco. Its habitat vanished under roads and houses but the Xerces blue is not forgotten. Its name is honoured and proclaimed by the Xerces Society, a voluntary body that is doing its best to ensure that no other American butterfly goes the same way. The butterfly itself is preserved as pinned specimens in museums.

The other butterfly extinctions were much less consequential. Japan recently lost a species of holly blue which were confined to the volcanic Ogasawara islands. South Africa has lost a couple: the Mbashe river buff and Morant's blue. Interestingly all four extinct butterflies belong to the same family, the Lycaenidae, the blues, coppers and hairstreaks. Their loss has been noted because the butterflies of North America, Japan and South Africa are relatively well-known and well-studied. No one knows what is happening to the butterflies of, say, Brazil, Angola or the Democratic Republic of the Congo.

In mainland Europe there have been no butterfly extinctions at all, though several are believed to be in imminent danger through habitat loss and climate change. As usual it is the small island endemic species which are the most threatened. The Madeiran large white was last seen in 1977. Its loss is attributed to parasites introduced to the island by imported cabbage white butterflies. Madeiran butterflies seem to be particularly vulnerable to outside influences. Also at risk are its distinctive forms of the speckled wood and the Cleopatra, both now recognised as full species. Their decline is attributed to the loss of the island's special Laurisilva forests, an example of a previously widespread type of laurel among which many unique species existed.

Just three butterflies have received international protection. One of these is the Queen Alexandra's birdwing of New Guinea. It is beautiful – the male is as iridescent as any hummingbird – and it is endangered, but then, so are many other butterflies. The reason it was singled out for top-tier protection is that it is the largest butterfly in the world. Many collectors would love to own a pair, and if it was on sale, demand would probably exceed supply. But the butterfly's rarity is due not to collecting so much as habitat destruction: the felling of indigenous forest to make way for palm oil plantations: even international protection does not prevent that. The eruption of a volcano in the 1950s also destroyed large areas of butterfly forest. Curiously enough, the Queen Alexandra's birdwing's commercial value – the black market price for a pair is around $US 8–10,000 – may be the saving of it. Income from the sale of valuable butterflies is an incentive for preservation; it means that people will have a reason to care about them. Local people are now rearing them, and related butterflies, from eggs, and even cultivating the vines on which their caterpillars feed. Income also encourages research, and there is now a captive breeding laboratory in New Guinea for endangered birdwing butterflies, sited, by another irony, smack in the middle of a palm oil plantation. As with so many endangered birds and mammals, captive rearing may yet save endangered birdwings. Of course, without sufficient old forest habitat to sustain them, the best the future can offer the likes of Queen Alexandra's birdwing will be butterfly farms.

Butterflies are not the hardest hit group of invertebrates, however. That's land snails. In terms of the proportion of species that are extinct or close to becoming so, land snails are the most threatened group of animals on earth. One analysis suggests that of all recent animal extinctions in the world, a full quarter were species of land snails. Another pushes that number up to 42 per cent when freshwater snails were also included. The extinct snails are mainly ocean island species. How they got to these remote islands one can only speculate: perhaps their ancestors floated there on driftwood or vegetation. Once established, and finding the islands to their liking, the snails evolved into a dazzling array of different though closely related species. Of the 400-odd species currently listed as endangered or vulnerable, 65 per cent of them are confined to islands, and often to just one island or island group. St Helena, for instance, has lost at least eighteen species of mollusc. Mauritius has lost at least thirteen. Of the fifty-four species identified on the IUCN's European Red List, forty-four are confined to islands in the North Atlantic or the Mediterranean.

The capital of snail extinction, the smoking crater so to speak, is Hawaii. On New Year's Day 2019, the island mourned the passing of George, a prettily banded tree snail called *Achatinella apexfulva* (it means 'yellow-tipped little snail'). George was born in a research facility and lived in a temperature-controlled box inside a modular trailer known as the love shack, because encouraging endangered snails to breed was its function. George was left-handed in that his coiled shell turned anti-clockwise ('sinistral'). He was named after the better-known Lonesome George, the last surviving giant tortoise on Pinta Island in the Galapagos, who died in 2012, aged 102. Snail George's ancestors lived high up in the trees of Hawaii's humid mountain forests and scrublands, subsisting by night on mildew. They rarely moved from the tree on where their mother had laid her eggs (if we can use the feminine for a species which is hermaphrodite). George and his fellow tree snails were long-lived, for invertebrates, with a correspondingly low fertility rate. George was about fourteen when he died, possibly of old age.

Historically, Hawaii was home to some of the most beautiful snails on earth. Smooth as alabaster, pellucid as amber, variously banded or striped, and every colour of the rainbow, their shells were eagerly collected to make *lei*, garlands or necklaces made by threading together their little shells. That alone may have stressed the populations of snails that were slow to replace their numbers, but what happened next is enough to make you bang your head against the wall (either that, or laugh out loud at the unintentional consequences of ill-considered actions). In the 1930s, there was a brief vogue among Japanese immigrants for keeping giant snails in their gardens, as exotic pets and also as a living larder, for edible pets are very useful in times of hardship. The species they chose was the biggest land snail on earth, *Achatina achatina*, the Giant African Snail, a heavy handful of a mollusc reaching the size of a rugby ball, and with an appetite to match. On its home ground in Africa, the numbers of *Achatina* are controlled by native predators and parasites. But released in similarly humid places else-where in the tropics, without those predators or parasites, it becomes one of the most invasive species on earth. In Hawaii, for example, the pet snails promptly ate most of the garden produce, and then moved into crop fields where they lived munching happily the tender leaves of papaya, banana and nut-trees, but not refusing crops grown for local consumption such as corn, beans and potatoes. For an extra bit of protein, it also ate other snails.

Naturally enough, everyone wondered what on earth to do about it, for once the stable door is opened, invasive species can become very difficult and expensive to control. Then, in the 1950s, someone had a brainwave: let's do it the biological way! In Florida, there is a snail with an attractive rosy shell called *Euglandina*, also called the wolf or cannibal snail because of its enormous appetite for chasing down and devouring other snails. It would eat the troublesome giant snails! Wouldn't it? *Euglandina* was duly released in Hawaii, and – guess what – it proved to be invasive too. Worse, instead of eating the giant snails, as expected, it turned instead on the native Hawaiian tree snails. Tiny, slow-moving snails, the size of a fingertip, made easier prey than ponderous giants the size of rugby balls. And so Hawaii's multitude of tree snails began to vanish from the forests. Of the forty-one species in the genus, *Achatinella*, for example, only thirteen still survive. And they are probably doomed.

To hasten their demise, another predator appeared in the 1970s, and by another foolish introduction. It took the form of Jackson's chameleon, a popular pet that subsists on snails, which it munches like a snack, shell and all. And, like the giant snails, chameleons set loose in gardens found the outback even more to their liking and so escaped into the forest. Hawaii has no native land reptiles and so its native snails had no defences against this one, just as they had none against ferocious wolf snails from Florida. One way or another, by tipping the balance of nature in a vulnerable island fauna, humankind has ensured the extinction of most, if not all, of the pretty shells once proudly worn as good-luck garlands.

There may be some connection with the loss of the tree snails and another unwanted introduction in the form of a pestilent fungus. It is killing off Hawaii's native Ohi'a trees in the same way that imported fungi are killing off Europe's ash trees. There is a suggestion that the tree snails, being grazers of fungi, might well have prevented this disease from spreading out of control. If so, it's yet another irony . . . When will we learn?

Snails aside, recorded invertebrate extinctions are fairly minimal. Anyone looking at IUCN's list might well wonder what the problem is. And yet a peer-reviewed study reported in the *Guardian* newspaper (Carrington 2019) confidently predicted the loss of half a million insect species over the next eighty years and leading to the possible collapse of ecosystems worldwide. The evidence, based on the known losses of insect biodiversity in study areas across the world, and on predictive models based on that data, made the headlines. The same statistics are repeated over and over on TV programmes about the mess we are making of our perfect planet. All the same, just as a detective looking for evidence of a murder might ask 'where are the bodies', so, we might legitimately ask, where and what are these doomed half-million? What are their names? The answer, I think, is that most of them have no names, and most of them never will. There are far more undiscovered, undescribed invertebrates out there than known ones. If some of them perish, they will do so quietly, unnoticed and unmourned: the 'Apocalypse Unseen', as the environment journalist Michael McCarthy (2016) dubbed it, or the 'Ecological Armageddon', as defined by the bee expert Dave Goulson (2017). Perhaps there is room for scepticism. The scientists who devise these predictive models are not necessarily good field naturalists. They might well not know one species from another. Their world is one of data acquisition and analysis, and so species become abstractions and animals morph into data. It's bio-maths, based on probability (though in some sense also based on data from the field). Such statistics are curiously theoretical and abstract. For it is hard to have feelings about mere figures.

Butterflies are the best studied insects on earth and yet, as we've seen, the names of globally extinct butterflies would fit on the back of a postage stamp. And because new species of butterfly are being described every year, the world list is growing, not shrinking. All the same, there is a growing body of evidence to suggest that this may be an illusion.

The famous zoologist and palaeontologist Richard Leakey predicted, in 1995, that by the time the world's tropical forests have been reduced to 10 per cent of their original extent, they would have lost up to half their species. No one knows how many species that would be, but it would probably be in the millions. Well, it seems that we have already reached what we might call the Leakey Point. In 2015, the world resource of these forests, the most biodiverse habitats on earth, had fallen to below 4 billion hectares for the first time. Some 8 million hectares per year continue to be lost (a slight reduction on the 10 million acres per year in the 1990s). The original extent of forest, before we started felling it, was around 400 billion hectares. That means there is now less than 10 per cent of tropical forest left, and that has been cut into countless fragments, like shrapnel from a bomb. How many species have been lost? No one knows. Was Leakey right? Again, no one knows.

Another straw picked up by journalists was from a paper in the journal *Biological Conservation* (Sanchez-Bayo 2019) which reviewed the results of seventy-three studies of insect decline throughout the world. The study concluded, gloomily, that insects are declining everywhere, that nearly half the insect species on earth are in decline, and that this decline is continuing inexorably at a rate of 2.5 per cent per year. This conclusion was an interpolation of evidence based mainly on attractive or economically important insects such as butterflies and bees. The paper points to agrochemicals as the villain, especially the neonics routinely used worldwide on crops (they are now banned for most purposes in the European Union and Britain). In the 1960s, thanks largely to American marine biologist Rachel Carson's warning in *Silent Spring*, DDT was banned, but we are still fighting those battles with different poisons (and Carson is still vilified by the industry).

Perhaps the most troubling evidence of all is a study of insect monitoring across nature reserves in Germany using Malaise traps, tent-like contraptions that catch flying insects (Goulson 2017). At the start of the project in 1989 these traps were catching an average of nine grams of insects daily. A quarter-century later, in 2014, they were catching just two grams. The data indicates that the biomass of flying insects in Germany had fallen by 80 per cent during that time. And bear in mind that these traps were sited on *nature reserves*.

In ordinary countryside the decline was presumably even

worse. Ecologists refer to arable-farmed countryside as a 'population sink'; in other words, flying insects that leave their habitat ghettos to pollinate the crops generally don't come back. Land use, and pesticide use, is basically the same in Germany as in Britain, France and Poland, and so it would be extraordinary if Northern Europe as a whole did not share equivalent levels of insect decline. In fact such evidence as there is suggests that this is indeed the case. Farmland birds that depend on insects to feed their chicks, such as partridges, have uniformly declined despite great efforts and subsidies devised to conserve them. And, as everyone knows, we no longer see swarms of insects in the headlights, and nor do insects come to lighted windows in the same numbers as before. Moth traps and other insect trap records tell the same story, one of decline in which some once common species seem to have crashed.

Is there any point in reiterating what would happen to a world without insects? It would be Doomsday. In E. O. Wilson's words, 'the environment will collapse into chaos'. Flowers would set no seed, birds and bats would starve, dung would not be recycled, and – hello! – here come the survivors, crop pests with no predators to keep them in check. In this horror-world the extinction of individual species would be a detail. What matters more is overall numbers, the extinction of abundance. Without the swarms, the outlook for Wilson's 'little things that run the world' looks bleak.

Five rhinos

As an icon of ancient life how can you beat the rhinoceros? The second largest land animal on earth has a defiantly prehistoric look. You think, that is what the great beasts of the distant past must have been like, heavy, horny, slow witted, and with a confident air that no other animal is going to touch you. Rhinos truly are an ancient line. Ten million years ago there were far more species than the current five, and they were found in temperate, even polar, latitudes as well as the tropics. The biggest bygone rhinoceros approached an elephant in size while the smallest was not much bigger than a pig (the indricotheres of the Oligocene, a hornless offshoot of the rhino line, were even bigger than elephants).

I have encountered rhinos in the wild, though I don't have any dramatic stories to tell. I was surprised at how easy it was. The first was a black rhino and its calf at a waterhole in the Etosha National Park in Namibia. They just wandered over while we were sitting there, safe behind a concrete barrier, as if they were entering the ring at the circus. As far as I remember, they stood stock still for about twenty minutes until it was me that left: I went inside for a beer feeling I'd seen enough! The adult black rhino which we encountered while driving in South Africa's Kruger National Park was far more skittish: it sniffed the air, spotted us and promptly crashed off into the bush with surprising speed. *Run, rhino, run!* The rhinos of Kruger have been cruelly poached. They say that, as their numbers have fallen, black rhinos have become more solitary. They just want to be left alone.

As for the larger white rhino, a placid grazer usually found in small family groups, they, too, were easy enough to locate, quietly munching away in the middle distance. They looked much the same at Kruger as they did when I'd seen them in the Cotswold Wildlife Park, although admittedly they were in a more impressive setting. Poachers were killing a thousand African rhinos every year, but these Etosha and Kruger beasts had managed to hang onto their horns – but then we saw them in one of the busier parts of the parks. In the Kruger, most of the poaching is done in the places which few tourists manage to reach, close to the Mozambique border.

As everyone knows, the rhinos of the world face a bleak future. Rhino horn, chopped from the living animal, smuggled to the Far East and consumed in ground-down form, is one of the most valuable substances on earth. It is worth more than gold dust, and the animals are being slaughtered for it. Rhino-horn is supposed to have miraculous properties, whether medicinal, or as an aphrodisiac, or – and this is a new one – a hangover cure. Some even snort the stuff like cocaine. People, mainly in China and Vietnam, insist that rhino horn is a fabulous material – it is a cultural given – and a growing number are willing and able to pay fantastic prices for it (the demand increases while the supply diminishes, and so the price goes up, making poaching all the more tempting). A single rhino horn can change the life of a poor poacher in Africa, though no doubt most of the profit goes to the middlemen who run the trade. All the killing is illegal. Rhinos are protected animals everywhere in the world. Trophy hunting, once the main threat, is no longer of much significance, but wherever there are rhinos there are now poachers, and poaching had reached an all-time high by the first two decades of the twenty-first century. At the present rate it is unsustainable. In Africa, at least, more rhinos are being killed than nature can replace. And, like drugs, the trade is highly organised and seemingly unstoppable.

Rhino poaching is unbearably cruel. These days, poachers carry sophisticated technology to track down animals. They commonly use a tranquilliser dart to bring down the beast, and then, while it lies unconscious but still alive, they hack off its horns with a chainsaw, mutilating the animal's head in the process. The shorn rhino bleeds to death, slowly and painfully. Modern poachers go into the bush well-armed. A friend who travels in remote parts of sub-Saharan Africa once told me that much of the horn trade there was (I use the past tense because the rhino is now extinct in that area) carried out by a relatively small band of well-armed poachers based in Sudan. Why can't they take them out? I asked, naively. 'Because they are very good at taking out the people trying to take them out,' he replied. Some 320 wildlife rangers were killed on duty between 2004 and 2018. They have to act within the law: the poachers do not.

There are just five species of rhinoceros, the scant survivors of the world's megafauna. The African species are the best known since they are in zoos and wildlife parks the world over, and star in wildlife films. The black rhino is the smaller of the two Africans and is a browser. The larger, white or square-lipped rhino is a grazer of different habits that enable the two to coexist. They are poorly named, for both are greyish, and in any case commonly coated in mud. South Africa has more rhinos than any other country – roughly 20,000 whites, up from a few hundred in 1970, and about 4,000 blacks, down from about 16,000 over the same period.

While Africa has two species of rhinoceros, the black and the white, Asia has three: the Indian, Sumatran and Javan rhinos. The Indian rhino, the second largest species after the white rhino, is now restricted to a few places in northern India, Nepal and Bhutan. Only around 2,200 animals survive in the wild, many of them concentrated in just one national park. The Sumatran rhino, the smallest species, was once widespread in Southeast Asia but is now confined to a few scattered populations in Sumatra and Borneo. There may be as few as eighty animals left in the wild. The Javan rhino is rarer still. Once similarly widespread, it is now confined to the western tip of Java where perhaps as few as sixty animals, maybe even fewer, still lurk in the dense forest. It has been called the rarest large animal on earth. It was commoner once. The thirteenth-century merchant-explorer Marco Polo saw some but thought they must be unicorns. The Javan rhino probably owes its survival to the eruption of Krakatoa in 1883. Its peninsula was devastated by the volcano; the human inhabitants fled, never to return in numbers. Jungle soon cloaked the abandoned land and that suited the rhino. It is rarely encountered, and if it sees a human the Javan rhino hares off in a thoroughly understandable display of caution.

It is hard to be optimistic about the future of any of the five rhinos, though, in a sense, we need to be. Like the great whales, the solution to the problem is simple enough. Stop killing them. There is enough space left in the larger national parks to support sustainable populations of at least the African species and maybe the Indian one too. Unfortunately poaching seems unstoppable. Even where there are local successes, the poachers can move on somewhere else. Money drives everything. Conservation remains a wish, not the actuality.

The group name for the rhinoceros is a *crash*. As I discovered during my brief encounter in the Kruger, black rhinos do make a crashing noise as they burst through the bush, but the word is apt in another sense. Rhino numbers have indeed crashed. As Mark Carwardine, the zoologist, puts it, 'The scary thing is that in my lifetime, 95 per cent of the world's rhinos have been killed.'

We are where we are because of what has happened in our own lifetimes. Can we yet save the rhino? Predictions of the future of species generally lean heavily on hope, for reasons buried in human psychology. If we lack hope, we might as well give up now. The bleak truth that is staring us in the face may be unpalatable.

All the same, a few facts might help. Let's look at it this way. Yes, there are only five species of rhinoceros on earth, but within each species lies a range of variation, genetic lines that differ in various anatomical features. All these forms are geographically separated and do not interbreed. And at that sub-specific level there are not five kinds of rhino on earth but eleven: two different forms of black and white rhinos, and three each for the Sumatran and Javan rhinos, making ten in all, plus the Indian rhinoceros. And here's the rub: of those eleven forms, *no fewer than six* are extinct, or functionally extinct, hunted and poached out of existence. More than half of the world's biodiversity of rhinos has been wiped out in our lifetimes.

Of the vanishingly rare Sumatran rhino, the last individuals of the distinct northern form were seen in the 1990s. The smaller Borneo form was also believed to be extinct until an individual was spotted in 2016, the first in forty years. Numbers there are almost certainly too low to sustain the species. On Sumatra, by far the largest population of some 500 rhinos was wiped out by poachers even though the animals lived inside a supposedly protected national park. That left just four scattered populations, one of which may already have perished. Some forty Sumatran rhinos have been captured and displayed in the world's zoos, but most of them soon died and none have bred. Not looking good is it?

Of the Javan rhino, the distinct form native to India had died out by 1910; its cousin in Malaya had gone by 1922, and in Sumatra by 1945. A second subspecies inhabited the forests of Vietnam. It was thought to have died out during the forest destruction of the Vietnam War until a small population was rediscovered there, within a supposedly protected national park. But Vietnam is one of the centres of the horn trade and by 2010, poachers had killed those too. That left only the population at Ujung Kulon on the western tip of Java (hence that misleading name, the Javan rhino). Conservation geneticists believe that a population of a hundred animals would be sufficient to sustain this species and avoid inbreeding. Current estimates of their numbers vary between eighty and thirty. There are no Javan rhinos in captivity. The last captive animal, called Samson, died in 2018.

In North Africa the original four sub-species have shrunk to just two. The western form of the black rhino used to inhabit the savannah south of the Sahara desert. Heavily hunted, it recovered somewhat after legal protection in the 1930s. But poaching did for it in the end. The figures for the years around the turn of the twentieth century sound like a countdown: 'hundreds' still roamed the sub-Saharan bush in the 1980s; but only ten animals were seen in 2000, just five in 2001. The last sighting of an authentic western black was in Cameroon in 2006. It was declared extinct in 2011. There are no animals in captivity; those rounded up for a segregated life behind a fence in 1988 all died. And that was the end of *Diceros bicornis longipes*.

The northern white rhino also stands on the very brink of departure. It differs from its southern relative in being smaller (but still enormous), with longer limbs and a straighter back, and with smaller, lower-crowned teeth. Historically, its range was restricted to the area where Sudan, Uganda, the Congo and the Central African Republic all meet – though in prehistory its range was much wider. Its demise through relentless poaching is ill-documented for this was, and is, a remote and unstable area. There were still around 500 northern whites in the 1970s but that number had shrunk to only 15 a decade later. It increased slightly to 32 animals in the 1990s, but to no avail. Poachers killed them all. The last known animal in the wild was found, dead and de-horned, in 2008. My *Collins Field Guide to the Larger Mammals of Africa*, published in 1970, tells me that the northern white rhino was 'well protected'. It wasn't.

The last surviving representatives of the northern white line lived in zoos, mainly at San Diego and in the Czech Republic. But rhinos seldom reach their full span in captivity. They were all given names. San Diego Zoo was home to Dinka and Bill, Lucy, Joyce and Nadi, and Angalifu, a fine old forty-four-year-old male, who died in 2014. The last of them was Nola, a female who died in 2015. Dvur Kralove Safari Park, a specialist centre for African fauna in the Czech Republic, remembers Nasima, Saut, Nuri, Nesari, Nasi, Nabire and Suni, and nearly all of them dead before their time. The last male northern white, called Sudan, after his country of origin, had to be put down in 2018, aged forty-five, when his wounds would not heal. With him vanished all hope of the continuance of his kind, at least by natural means.

That left just two animals, both female. The last northern

white rhinos on earth are Najin, now aged thirty, and her daughter, Fatu, aged nineteen. They are both tame animals, habituated to humankind. They were born in the Czech Republic but were transferred to native soil at the Ol Pejeta Conservancy in Kenya, in 2009. Now heavily guarded around the clock, behind a high fence, they are the most famous rhinos in the world, placid animal celebrities that treat their visitors with the same mild curiosity. But while Najin still seems healthy, her daughter, Fatu, has weak legs and no lining to her uterus. Although both had mated with their now dead male suitors, Sudan and Suni, there were no pregnancies. As Najin's keeper, James Mwenda, says, 'When [she] passes away, she will leave her daughter alone forever.' Unable to complete her biological function on earth, all she can do is to live and then die.

Scientists have joined the increasingly desperate efforts to save the northern white. Sperm had been extracted from the last surviving males but it proved weak and of poor quality and attempts at artificial insemination also failed. Besides which neither Fatu nor Najin will ever be mothers, the one being too weak, the other too old. The only remaining practical possibility is to extract eggs from the two with a view to making an embryo in the lab and nourishing it inside the only proxy mother available, a captive *southern* white rhino. The project is fraught with danger for the rhinos because their anatomy requires the insertion of a long needle up through the anus and into the ovary to suck out an egg. This delicate operation, carried out on the unconscious animal within the maximum safe period of two hours, risks puncturing a major artery and killing her. Fortunately the rhino survived the operation and the lab people got their eggs. They duly managed to create three embryos for the world's first 'test-tube' rhinoceros. At the time of writing these were being stored in liquid nitrogen awaiting a suitable proxy mother. And you have to think, kudos to the patience, dedication and skill of the white coats, of course. But what kind of solution is *that*?

A television documentary about this project was optimistically titled *Return of the African Titans*. The series on endangered animals, presented by Mark Carwardine and Stephen Fry, had a more appropriate name: *Last Chance To See*.

In the Kruger National Park, which has by far the largest numbers of southern white rhino, tougher sentences and better patrolling did reduce the number of animals killed. But poaching has correspondingly increased in some national parks, and in countries such as Botswana and Zambia. Besides which effective protection is being hamstrung by legal hurdles. The local court at Kruger, which had been very effective at prosecuting poachers, has been disbanded, and arrest rates have declined again. And the chief of police wants to ban the carrying of semi-automatic weapons by rangers supposedly on safety grounds. If so, the rangers could find themselves outgunned by the poachers.

That was the position at the end of 2020. That same year, the IUCN drew a formal line under the northern white rhinoceros. It is now officially 'critically endangered (possibly extinct in the wild)'. Coincidentally, new research suggests that the northern white might have been sufficiently distinct from its southern relative to be a distinct species, much as the African forest elephant has been split from what is now called the savanna elephant. If so, it is a wicked irony that its significance was discovered at the very brink of the grave. We could call it a *post-mortem* species: a retrospective distinction, like the medals awarded to departed war heroes.

How would we feel if the only rhinos left in the world were those living, in every sense, truncated lives in zoos, safari parks and farms? The lesson we would reluctantly have to take is that when there is a large price on an animal's head, there is no longer any safe place for it on earth. Only captivity, far away from their home range, would remain. A large number of well-motivated, even noble people are doing their level best to prevent this from happening. We must hope they succeed, but the line on the graph suggests a different outcome. Plotted numbers against time, the line rolls downwards with the remorseless insistence of mathematics.*

Helpless, the world watches and waits.

---

* Every species in decline will have its own 'line on the graph' when plotted as numbers against time. Of course, only species with monitoring data will have a known line-trajectory, as opposed to a theoretical one. But a continued downward trend indicates a journey towards extinction, whenever that eventuality will actually come. It reminds us that extinction is not only a moment, the last moment in the life of a species, but a process. The line represents hard truth, buck it as we might try. If I ever write another book about extinction, I'm going to call it *The Extinction Line*, the roadmap of species in free-fall.

Chapter 3

Extinction on one island: the British experience

Here we are ... filled to exuberance with our new understanding of kinship to all the family of life, and here we are, still nineteenth-century man, walking boot-shod over the open face of nature, subjugating and civilising it. And we cannot stop this controlling, unless we vanish under the hill ourselves. If there were such a thing as a world mind, it should crack over this.

Lewis Thomas (1974), 'Natural Man', in *The Lives of a Cell. Notes of a Biology Watcher.*

We know when the thylacine or Tasmanian wolf died out. It was on 6 September 1936, and he had been the last of his kind for six years. He lived in Hobart Zoo and his name was Benjamin. We can still watch Benjamin on YouTube, pacing restlessly up and down, occasionally yawning in boredom. As a species, the thylacine had lasted about two million years and, for the past 50,000 of that, it had been in decline.

In Britain, we thought we knew exactly where and when the last of the distinctive British race of the large copper butterfly died out. It was during the summer of 1851, at Bottisham Fen near Cambridge, when the very last butterfly was caught by a Mr Wagstaff of Chippenham (or then, again, maybe not: Peter Andrews (2020), digging into the evidence, discovered that the last reliably reported large coppers actually flew a few years earlier, in 1847 or 1848, at Holme Fen, near Peterborough). It is seldom that one can date an extinction so closely, let alone pick on someone to blame. Extinction, national or global, is usually retrospective, deduced from a continued absence. As I say, most species disappear quietly. They join a slow, barely noticed leakage of biodiversity, one drop after the next in the ocean of eternity.

Take one of Britain's 500 or so 'declared' lost species, the horned dung beetle, *Copris lunaris* (a poetic name, as Latin names often are: it means 'dung beneath the moon'). A chunky, good-looking beetle (for a beetle), shining in the moonlight as if made of polished ebony, it is named after a little rhino-like horn sticking out at the front end. This made it easy to recognise, and hence, when the beetle could no longer be found in its old haunts, on Box Hill, or the downs near Guildford, someone decided to check the database. It seemed that the last time anyone had found *Copris lunaris* in Britain was in fact 1974 (it still occurs on Jersey). However it continued to be docketed as 'critically endangered (possibly extinct)' on the off chance that someone might find it, still rolling little pellets of dung into its burrow by starlight. Apparently when stressed by circumstances, *Copris* stridulates like a grasshopper. Perhaps, then, if anyone had witnessed its last moments on British soil, they might have heard the high-frequency buzz of a beetle in distress. Why did it die out? When you are down to the last few colonies, survival can be tripped by random events: increased shade, a summer's overgrazing, perhaps toxins in the dung left over from veterinary treatments. So, we'll probably never know. And no lessons learned, except that dung beetles generally are being hit hard.

Extinction has a way of taking us by surprise. Predictions are often wildly off. A half-century ago, the British butterfly thought least likely to succeed was the silver-spotted skipper, a species with rather fussy habitat needs. Far from dying out, it has seen the biggest proportionate increase of any British butterfly, admittedly with the help of some bespoke management on nature reserves. About twenty years ago, the same people were shaking their heads over the prospects of the little Duke of Burgundy, whose status seemed to be equally parlous. That, too, has 'defied augury' (as Shakespeare puts it in *Hamlet*), partly because it thrives best on chalk downs which are scrubbing up, as too many of them are, and partly because it is a quiet sort of butterfly that is easily overlooked. On the other hand, they predicted a sunny future for the Glanville fritillary and the Lulworth skipper as our summers grew longer, warmer and balmier. That hasn't happened either – and neither, at least in some years, have our summers.

Sometimes a species is believed to be endangered when, in truth, it is only because we lack the knowledge of how and where to find it. In the 1990s, I was asked to investigate the status of a large and colourful fungus called the devil's bolete, *Rubroboletus satanas*. It was supposed to be critically endangered, possibly nearly extinct, since there were only two recent records in the national database. Since these are great fat fungi the shape of staddlestones, and relatively easy to recognise, it seemed that an absence of records indicated an absence of fungi. But if so, why had it gone? What ailed the bolete?

The answer, which only took a little research, was that it was the records that were missing, not the bolete. Bolete specialists tend to work in isolation, each with their own, private database. Moreover the devil's bolete often appears well before the recording season gets underway. It seems to need just the right combination of warmth and humidity to fruit in Britain, and in cool or too dry summers it may not appear at all. When they put 'my' devil's bolete on a stamp that year, one of a set of endangered species, I felt proper pride. But it would be cynical to claim that I had 'saved' it. It was never endangered in the first place.

I once wrote a paper on a plant that really is endangered in Britain and, in fact, probably doomed. The alpine or blue sow-thistle, *Cicerbita alpina*, is our tallest and in many ways oddest mountain plant. In Europe, it grows under the shelter of trees, especially birch or spruce, in the lee of mountain slopes. In the right environment, it can grow as densely as bluebells, its stems, with their terminal clusters of pale blue flowers, towering over a bed of lush, arrow-shaped leaves. Open woodland is the natural home for such a plant, sheltered from the wind and frost under a safety net of boughs. In Britain, though, it no longer grows under trees but rather on exposed north-facing cliffs, generally of granite, on the highest mountains of Aberdeen and Angus. As I found for myself, in my clambering days, the sow-thistle seems unsuited to such a vertiginous life. Flowering in late summer, the plants quickly become battered and bowed by wind and rain, rather like someone waiting in a bus queue during a downpour. If any plant could be said to look unhappy, crushed by misfortune, this is it.

The blue sow-thistle is stuck on its exposed ledges not out of choice but out of necessity. As a big, leafy plant, it would make a nutritious snack for any passing deer or sheep, and in Scottish conditions would not last five minutes unless it was flowering somewhere out of reach. Over the last century-and-a-half, the Scottish Highlands have been grazed flat. Unless kept out by a fence, the unnaturally large numbers of red deer will strip away all but the toughest plants. A mid-nineteenth-century *Flora of Angus* caught the very last time the sow-thistle grew in Scottish mountain woods, as it still does across the sea in Norway. So the places where it grows today are its default setting, the relatively few cliff ledges that are sufficiently broad and sheltered to accumulate deep soil and so offer it a lifeline.

Every species has its graph, numbers against time, a line running upwards or downwards or wiggling along horizontally. Healthy populations can withstand a bit of up and down. They bounce along the years, regulated by any number of imponderables – the weather, parasites, food, the state of the habitat. They dip and recover, recover and dip. Rare species are more vulnerable than common ones because they lack resilience. Since their numbers are low to start with, when circumstances turn against them the line may dip beyond any possibility of recovery. Of course some plants live longer than others, and the process may be so slow as to be beyond detection. Only long-term monitoring will reveal what is taking place under our noses.

In the case of the blue sow-thistle, a limited amount of monitoring (by myself, and fellow alpinists like Sandy Payne and John and Pamela Clarke) suggests just such a slow, inexorable decline. There seem to be only a hundred or so genetically distinct individual plants left in Scotland (each plant can put forth several flowering stems). For an out-crossing plant, this is probably well below the safety line. Low genetic variability condemns it to weakening disease resistance and probably accounts for its inability to set fertile seed except rarely. In the long-term, these circumstances spell likely extinction. Everything is against it. The chances of cross-pollination between the four extant populations, each many miles from the next (and together covering a space smaller than the average kitchen), are close to zero. The chances of it spreading beyond its beetling cliffs are also negligible for the over-grazing of the Highlands is growing worse, not better (some 300,000 red deer in 1980; half a million now). There is no sign that the plant has spread anywhere for at least a century; indeed, how could it without seeds?

For this reason the Royal Botanic Gardens in Edinburgh has begun cross-pollinating British plants with genetically healthier stock from Norway, to inject some new blood (all right, sap) into the failing Scottish stock. There has also been a limited amount of translocation of cultivated plants into places prepared for it on lower ground. Unfortunately all those plants got eaten by voles. Like so many of our rare plants, blue sow-thistle is easy to grow in gardens (voles permitting) and that might well be its future in Britain, as a garden plant, a wild garden inside a nature reserve perhaps, but still a garden where all the decisions will be taken by gardeners, not nature. If so, it would no longer, in any meaningful way, be a wild flower. That may well be the last stand of the blue sow-thistle in Britain, not growing bloody but unbowed on the great cliff of Lochnagar (with a lovely view of Balmoral) but under a net in a prepared cultivation bed, surrounded by vole traps.

Biodiversity's slow bleed

Island Britain is sometimes called one of the world's most nature-depleted countries. But if that is true it is because many of the dominant species were lost in the distant past. Most of the native megafauna – bear, elk, wild cattle, reindeer, wild horse – were driven out in prehistory. Wolf, lynx, beaver and wild boar survived into historic times but they too are only a distant memory, apart from the beaver which is being expensively reintroduced by rewilders, and the wild boar which has reintroduced itself after escaping from farms. The largest land animal left is the red deer, which was protected as game. Our largest land carnivore is the badger.

How many species have actually died out in poor, nature-depleted Britain? In 2010, the International Year of Biodiversity, Natural England (NE) published a report on *Lost Life: England's lost and threatened species* (because of devolution we are in the ridiculous position of discussing losses on only half the island of Britain, although the figures mostly hold good for the whole of the UK). By Natural England's reckoning we have lost about 500 (492) species of native plants, animals and invertebrates since 1800. Given that England has an estimated 55,000 wild species known to science, this represents a loss rate of less than one per cent, or just over two species per year. (Natural England's list excludes fungi, algae and marine invertebrates and so the figure is probably well short of the true total, which is unknown.) Compared with the recent record of most islands around the world, Britain has not done too badly. Our wildlife is, on the whole, fairly resilient. One way of looking at it is that the Sixth Extinction has barely begun here. Another is that it started a long time ago.

Few of the lost 500 are familiar. Many even lack a common name. And nearly all are still found elsewhere in the world, sometimes commonly, and in some cases just on the opposite side of the English Channel. The most famous globally extinct species on the list is the great auk, that flightless 'northern penguin', which probably last bred in the British Isles in the 1820s and was gone from the earth by 1844. Otherwise the best we can do by way of world extinctions is 'Mitten's beardless moss', which may not be a full species at all, and the tiny Ivell's sea anemone, *Edwardsia ivell*, no bigger than a fingerprint. The latter was discovered spreading its little tentacles at the bottom of Widewater Lagoon near Chichester, West Sussex, in 1973. It had vanished a decade later, and its short life was witnessed only by Richard Ivell and a few fellow specialists. Its habitat of a landlocked pool of brackish water is limited in extent and vulnerable to pollutants. By its disappearance, the Ivell's sea anemone is implicitly telling us something about the state of its environment.

Britain's most impressive lost beast is the northern right whale, harried to death by Basque whalers back in the nineteenth century, and now functionally extinct along the whole eastern seaboard of the North Atlantic. 'Functionally extinct' means that the population is no longer viable, even though the odd whale will visit us from time to time. It survives in small and depleted numbers on the far side of the ocean. At a pinch we could also include the world's largest animal, the blue whale, which occasionally strayed into British waters, but not since 1897 when the last 'British' blue was found dead or dying on the beach at Boscombe, Hampshire, after being struck by a ship. ('Children took great delight in climbing and sliding down its sun-bleached bones' reported the *Bournemouth Echo* that year.) The skeleton now dangling from the rafters in London's Natural History Museum is of a slightly earlier stranded blue whale from Wexford in Ireland. It is called 'Hope', the museum's 'symbol of humanity's power to shape a sustainable future'.

Whales apart, the only completely lost British mammal of the past two hundred years is the greater mouse-eared bat, which was never well-established, and whose two tiny colonies eventually dwindled down to a single, long-lived male. The wild cat – usually called the Scottish wildcat, since Scotland is or was its last refuge in Britain – was declared functionally extinct in 2018. A study of wildcat DNA indicated that all, or nearly all, surviving Scottish wildcats are hybrids, having bred with domestic moggies somewhere along the line. They are still wild cats, but not true-bred wildcats.

The black rat, too, may be functionally extinct since we deliberately wiped out the last sustainable colonies to protect nesting seabirds. Nobody mourns it and nobody is saying we should reintroduce it either. Good riddance seems to be the universal view, for which harshness one could perhaps attribute a lingering memory of the Black Death, a pandemic disease transmitted by rat fleas.

British avian extinctions are also relatively minor. Birds have wings and since the sea is no barrier to them, the return of temporarily lost birds such as the avocet or the osprey was always on the cards. Of those that did not return, or at least not under their own devices, the great bustard was shot to extinction by 1832; the Kentish plover left us in the 1930s; and so did the black tern which in any case only nested occasionally. More recently, we lost the red-backed shrike as a regular breeding (and familiar) species, although the odd pair still nests from time to time. The same is true of the wryneck, another once familiar bird of gardens and orchards. Possibly their decline is linked to that of large insects. The golden oriole has ceased to nest in its last English site (a nature reserve) for the time being, but it might try again. Set against these is a host of new breeding species, especially coastal and wetland birds: egrets, common crane, spoonbill, Cetti's warbler, Mediterranean gull . . . the gains outnumber the losses by more than two to one. Being so mobile, birds are more able to take advantage of such opportunities as are presented. It is when their food starts to run out, as with the falling numbers of Arctic tern and kittiwake, that the trouble begins. The same is true of some of our farmland birds; the monitoring scheme suggests that many are in trouble, again because of the falling food supply on today's intensively managed farms.

Of Britain's few native amphibians, the agile frog and moor frog disappeared at some uncertain date, for unknown reasons; they are known only from their bones. The native pool frog was long confused with the introduced edible frog, and no sooner had its real identity been revealed, in the 1990s, than the frog disobligingly died out. It has been reintroduced from Sweden, whose pool frogs are believed to share a similar genetic stock, but it is too early to call it a success. As for fish, we have lost the burbot (see p. xx), the sturgeon and the houting, and are perilously close to losing the vendace and the char.

Once we reach the invertebrates, the field broadens out. According to Natural England, we have lost: 18 butterflies, 89 moths, 64 beetles, 23 bees, at least 14 flies, and 12 wasps, among smaller numbers of lost ants, stoneflies, bugs, earwigs, snails and fleas. The report's designers ran their names beneath the text in a litany for the dead: apple bumblebee, minutest diving-beetle, marsh dagger, the many-lined, a brine shrimp . . . It has to be said that some of NE's figures are questionable. On my calculation, England (and Britain) has lost 5 butterflies, not 18; and I'm hard put to find 89 extinct moths either; I make it around 60.

Why does one species die out while another survives? In the case of the black-backed meadow ant, *Formica pratensis*, last noted in the Bournemouth area in 1987, it might have been chance and bad luck. Some of its former sites now lie beneath the streets of Bournemouth and Poole. As for the Chilterns mini-miner bee, *Andrena floricola*, of which only a single example was ever found, who can say? Many of our lost insects are known from only a handful of records, and most of those not recent. They are almost 'ghost species', flitting briefly into light and then out again, back into the unknown.

In one or two cases, a species went extinct because we found it a great nuisance and so did our best to exterminate it. Few will mourn the loss of the two warble-flies, whose apt Latin name is *Hypoderma*, the once pestilent 'gad-flies' of horses and cattle. Another case of good riddance might be the skin mite, *Psoroptes equi*, whose bites caused mange in horses. Modern chemical treatment dealt with it with remarkable ease. I suppose one could argue about whether we have the right to wipe out any species, claiming that every single one has the right to existence, but biting and disease-carrying ones have the least grip on our affections. You might think we could do without parasites too, but insect numbers are held in check partly by such 'nasty little critters'. Without them, a species' numbers might swell beyond the capacity of the environment to sustain it, and then they would all be dead, host and parasite alike. It's a harsh world out there, but it works.

What can be extrapolated from Natural England's list? It suggests that extinctions among insects, at least, are concentrated in south-east England. There are remarkably few in the north but the north is also less well-recorded. The south-east has the greatest natural biodiversity and it is where you would expect to find continental species clinging to our shores on the very edge of their natural range and so particularly vulnerable to shifts in their environment. Certain groups, such as butterflies, bees, stoneflies and dung beetles, seem to have a disproportionately large number of extinctions, but that may partly be because they are relatively popular and well recorded (it also seems to hold true worldwide).

Occasionally one spots a definite anomaly. Land snails have the highest rate of extinctions worldwide of any invertebrate, yet in Britain it seems as though we haven't lost a single one.

For most invertebrates, extinction can only be probable rather than certain. Even with Britain's small army of specialist recording groups, how sure can we be that an ant or a mite or a woodlouse has really vanished? Rarities are hard to detect and one tends to look for them in their known haunts rather than striking out elsewhere. Invertebrates do have a habit of resurfacing again after a prolonged apparent absence, and in fact labelling a species as 'probably extinct' can act as a spur to try and find it again. A good example is the family of oil-beetles, the Meloidae. There are ten species of these large, violet-black beetles on the British list, but of those six lacked recent records and on that basis were assumed to be nationally extinct. But to identify beetles you need a good field guide, and that was the trouble: the missing oil-beetles were not illustrated in books or websites then available. Then, in 2006, one of them, the short-necked oil-beetle, *Meloe brevicollis*, was found alive and well, near Prawle Point in Devon, and subsequently elsewhere, in southern England, as well as in Ireland, Scotland and Wales. It was not only not extinct but actually widespread! A second of the missing, *Meloe mediterraneus*, subsequently turned up, also at Prawle Point. Finally, in 2010, the related flame-shouldered blister-beetle, *Sitaris muralis*, was rediscovered in the New Forest, and later in Wales and other places in England. That left just three species on the extinct list, one of which has not been seen since 1882 and may have been a stray. Much of this success in tracking down missing beetles is due to the efforts of the charity Buglife, which has shone a much-needed spotlight on less popular invertebrates.

A more recent leap from the grave was made by one of Britain's largest spiders, the great fox spider, *Alopecosa fabrilis*. Furry and reddish-brown, and up to two inches across its outspread legs, this spectacular spider makes a burrow in bare sand, often cunningly concealed beneath a stone. It emerges only at night to hunt down its prey with the help of two great staring black eyes, like dark headlamps (six smaller eyes are dispersed about its head). Despite its fearsome size, no one had seen a great fox spider in Britain since 1999. A local enthusiast, Mike Waite, was determined to rediscover it and spent some time searching the military training area of Surrey at night with a torch. He was rewarded by the sight of great hairy spiders running about, several males and a female, plus some baby spiderlings. The great fox was alive and well and making babies under the guns of the British army! This link with the military was probably no accident. The spider enjoyed a good war in 1939–45 when its habitats were used for training, thus ensuring plenty of bare, disturbed ground dispersed among the heather. Like many rare invertebrates and plants, it relies on regular surface disturbance. Putting a fence and a 'don't disturb' sign around its home would simply destroy its habitat.

Are we winning?

As we have seen, extinction is an instant but decline is slow. In some cases it might continue over hundreds, even thousands of years, and the moment at which a species hits the danger point is in the eye of the beholder. DNA studies suggest that one genetic group of woolly mammoths died out about 35,000 years ago, and another around 25,000 years later, while the last animals pegged out on their desolate Arctic island only 4,000 years ago. Among British species, the wolf was probably in steady decline for at least a thousand years. The numbers of lady's slipper orchid were falling from the moment it was discovered, 400 years ago.

That is what is so concerning about Natural England's report: not the number of actual extinctions but the much greater number of species in danger of meeting the same fate, in some cases possibly quite soon. NE singles out 943 threatened species in need of conservation action, implying that without such action they may die out. That includes all of England's whales and dolphins, all six native reptiles, and four out of the seven native amphibians. Add to that nearly half our freshwater fish, a third of our mammals, a third of our bumblebees and butterflies and 142 species of moth, and the overall sense is of biodiversity in trouble. Some of these species are well-known and well-loved: red squirrel, wildcat, eel, corncrake, water vole, pearl mussel; and in the sea, cod, haddock and skate. Imagine the sea without cod! On the Grand Banks of Newfoundland, they couldn't, until it happened.

Britain's plants are not getting off lightly either. The endangered list includes 128 species of native vascular plant (that is, flowering plants plus conifers and ferns), 61 mosses, 22 liverworts, 99 lichens and 9 stoneworts (large freshwater algae that look a bit like seedling pines). Past generations would have been astonished to see on the list once common, everyday flowers like pheasant's-eye and cornflower. Ominously, Natural England notes that 412 – nearly half – of our threatened plants, animals and insects are known from fewer than five sites. They may be nearing the end point of a long and gradual decline. On the positive side, it claims that, for a few, the 'conservation targets' have already been met. Among the lucky ones are the pipistrelle bat, the pink meadowcap toadstool, the slender green feather-moss and the western ramping-fumitory – but much as one rejoices for the western ramping-fumitory, it was due to the chance discovery of new sites more than active conservation.

For the majority of the threatened species, the most urgent need is to find out more about them and their way of life. Without knowledge, conservation is blind. Sometimes the solution is pretty obvious but also unattainable. The northern bluefin tuna has been 'fished to the brink' and the perfect solution would be to stop fishing until it has had time to recover. But that goal, although widely agreed between governments, has proven hard to enforce. It may already be too late for the 'common' skate whose population in British and Irish waters have become so reduced as to be unsustainable; perhaps it is already 'functionally extinct'. In the case of red squirrels, since it is impossible to get rid of its introduced rival, the grey squirrel (though we keep trying), the solution is to restock places, especially islands, which the grey squirrel hasn't reached. For the water vole respite lies in the maintenance of its habitat and war to the death on its exterminator, the mink. For bats it lies partly in persuading householders that they are lovely and useful animals which we should feel proud to nurture in our roof-spaces.

Detailed accounts of what is happening to the wildlife of Britain and Ireland can be found in a succession of academic books. The first, published in 1974 as *The Changing Flora and Fauna of Britain*, saw habitat destruction, and especially the advance of modern, chemically dependent agriculture, as the main problem. At that time, people were also worried about air pollution, especially sulphur dioxide from coal-fuelled power stations which combines with water in the clouds to produce acid rain. Nature conservation was still relatively weak and under-funded, and species monitoring had barely begun (though a Biological Records Centre had been established in the 1960s and used a proto-type computer fed by punch cards). Most information back then was anecdotal, and for the less popular forms of life the authors admitted that they really hadn't the faintest idea what was going on.

A quarter-century later, in 2001, *The Changing Wildlife of Great Britain and Ireland* was published under the editor-ship of David Hawksworth, an eminent mycologist. There was now a good deal of information from various moni-toring projects, and action plans were in place for certain threatened species. Strands of hope were noted. The air was cleaner, and so were some rivers. The threat from acid rain had receded (much of our polluted air blew east, to Norway). There was greater public awareness and more support for green projects. But there were also new problems, or at least a realisation that they were problems. Native species faced increased competition from invasive imports, such as the New Zealand flatworm or the American signal crayfish or the American mink, stupidly released from fur farms by animal activists. Global warming, barely suspected in 1974, was making itself felt. And some of the authors sensed a growing disconnect between people and nature. There were no more school nature tables, and the science of taxonomy – the describing and naming of species – was falling into abeyance as a perceived Victorian anomaly in the age of DNA. In common with the rest of the developed world Britain was becoming increasingly urbanised, meaning fewer people could recognise everyday wild flowers and fewer boys brought home frog spawn or caterpillars in jam-jars. Birds, though, remained super popular.

A third prospectus of British wildlife with the distinctly pessimistic title of *Silent Summer* appeared in 2010, edited by Norman Maclean, a geneticist at Southampton University. By now it was clear that some species were declining more than others, and that flying insects were being particularly hard hit. Gone were the 'moth snowstorms', those blizzards of insects you used to see in your headlights on warm, still summer nights. Anglers lamented the loss of mayfly hatches on trout streams. DDT and its derivatives had been banned but their replacement systemic pesticides were even more deadly destroyers of insects, including useful ones like bees. There were knock-on effects from more frequent flooding, from groundwater abstraction to feed dry cities, from coastal erosion, and from the increasing pressures of development. The population of Britain rose by 15 per cent to 60 million between 1960 and 2010, with a further predicted rise to 75 million by 2030. Can 75 million people on an already crowded island coexist with a thriving wildlife? Some wildlife charities claim we can, but only if we all become green and thrifty and vegetarian. *Right.*

Most of the authors of these texts claimed to be confident we would sort things out; that behind every problem lurks an opportunity. Professor Maclean was one of the exceptions. He warned that we would have to live 'without insect abundance'; with fewer moths on the windowpane, fewer flies at a picnic, less to see on a country walk. He foresaw growing problems from chemicals, including plastics, and from light and noise pollution. Dark skies were retreating to the least populated parts of the island: what effect did this have on bats and moths? Motorways and high-speed railways operated in a continuous roar: could songbirds make themselves heard? Yet even Maclean believed that most species were 'sophisticated' enough to ride the storm. He foresaw relatively little national extinction. And on climate change it was still a matter of wait and see.

Climate change observations have come into their own only in the past decade (though it feels longer). In 2018, Trevor Beebee of the University of Sussex would write a whole book on the subject in a purely British context, on its effects, on the acceleration of change and on precisely which species are benefitting and which are not. The problem with climate change is that it is impossible to make realistic predictive models: there are too many variables. In the long term, we do not even know for sure whether Britain will get hotter or colder (though the warming of the world in general is a given). In these circumstances, we can only guess how wildlife will respond. To take one small instance, no one predicted the explosive increase in prickly lettuce, an untidy weed with dandelion-like flowers. Suddenly it is everywhere though no one knows how or why. On the other hand, modelling indicates, unsurprisingly, that some of our northern and Arctic plants will take a pasting. Dwarf willow, for instance, a mat-forming, mini-tree of mountain tops is set to lose half its range by 2080, along with all the species that depend on it. A comprehensive study of the wildlife of the Cairngorms (Shaw and Thompson 2006) calmly predicted that an increase of just one degree Centigrade would be enough to reduce the area of alpine or sub-alpine habitat by 90 per cent across the whole of Scotland. Since temperatures are likely to increase a good deal further within the present century, the national extinction of some of our cold-climate species seems possible: they include snow bunting, ptarmigan, mountain ringlet butterflies and some very pretty alpine flowers.

Britain's anglers will not be happy either at the predicted fate of salmon and trout, which depend on clean, well-oxygenated water and gravel beds to spawn. In the uplands, these conditions depend partly on snow-melt. Even widespread freshwater fish like chub and gudgeon are predicted to decline as water temperatures rise and dissolved oxygen levels fall. In the sea, cold-water fish like herring and sprat will retreat northwards to be replaced by species currently swimming around the coast of Portugal and Spain, such as horse mackerel, anchovy and sardine. All this will have knock-on effects on seabirds. Those that depend on sand-eels, such as puffin and kittiwake, will need to move on, probably northwards, or starve.

The tendency of these publications has been towards increased pessimism about the future, and I know that many leading British scientists and naturalists are indeed less optimistic than they were even ten years ago.* But the counter-flow is a vastly increased public concern about the environment and what we are doing to the world. Conservation needs hope, even, in fact especially, in circumstances that may look hopeless. We can learn from mistakes and do things better and convince ourselves that we are making progress. Nature-depleted we may be, but recent extinction in Britain has been relatively mild. Whether or not it stays that way is, in the end, up to us. We are in charge now, or so we believe.

Time's ratchet: losing wild flowers

Wild plants offer a sense of permanence. They are rooted to the spot, self-sufficient in their needs, and have bountiful seeds or berries that will take their genes to other places and continue the line. We see them flowering year after year in the same place, seemingly as durable as milestones. Trees used to define the boundaries of estates in the same way as woods or streams. They were living markers, green signposts and they outlived the owners of the land. Some trees have outlived entire dynasties.

---

* Derek Gow, the eminent rewilder, believes that 'there is no such thing as an ecosystem anymore. What you have is ragged threads of what once was, blowing in a broken window. I'm not optimistic at all. We are looking at a knackered, exhausted planet that is slipping away from us'.
  Derek Gow in an interview for *The Times*, 30 July 2021.

Hunting for flowers has long been popular in Britain, a passion reflected in the long run of 'county floras' dating back three hundred years. Hence the changing fortunes of wild plants have been followed in some detail. There are also national botanical atlases comparing past and present distributions – tremendous tomes published in 1962, 1968 and 2002 (with an 'Atlas 2020' on the way). On top of that there have been specialist atlases on ferns, aquatic plants and 'critical' species, as well as numerous monographs on particular families of plants. The British flora is often called the best studied in the whole world. In truth, few of us know whether that is really true, but we say so anyway, because we are British.

You might suppose then that it would be a simple enough matter to say how many wild flowers are extinct in Britain. In practice the number is elastic, depending on which plants you are counting. Do you, for instance, include long-established but non-native plants such as poppies? Do you include 'critical' species such as hawkweeds or brambles, which run into hundreds of varieties which only a few specialists can name? And what do you do about 'doubtful' species like Arctic bramble, for which there are only a few records or specimens, and those long ago. It might have existed once, there might have been some mistake or even fraud.

In fact, the closer you look, the more the facts tend to dissolve. Yes, there are nationally extinct British plants – probably every country in the world has its own list of lost flowers. Natural England came up with the figure of twenty for England alone; coincidently the same number as in my book, *Britain's Rare Flowers* (1999), though not the precise same species. The GB Red List (2005), on the other hand, includes just nine, plus four more that are 'extinct in the wild' (for they live on in seed banks and plant pots). None of the twenty (or nine, or thirteen) are well-known, and they have not been much missed. There are no trees on the list, nor any cherished blooms. Their names are re-assuringly obscure: Esthwaite waterweed, Davall's sedge, jagged chickweed, downy spurge. Only one species, thorow-wax, might have been familiar once. Appropriately enough it has been used in funeral wreaths. Several species once believed lost have been rediscovered (like the military orchid, see p.230). Only one per cent of the British flora has died out during the past two hundred years, which works out at just one species every twenty years.

What ailed our lost plants? Why them? Several were field weeds, and so may have been lost through agricultural advances. Weedkillers, factory-made fertiliser and crop-breeding have all made life difficult for flowers that used to grow among the corn. Change in the flow of the maritime currents may have accounted for our two lost seaside plants, cottonweed (which yet survives in Ireland) and purple spurge; with another climate-induced twist, they could yet return. Our single lost orchid, the summer lady's-tresses, might have been a victim of collecting. Anchored shallowly in boggy ground, it had a perhaps fatal defect: if you tried to pick it – and lots of collectors did – the whole plant comes up by the roots. But we know where it grew and those places would be too dry or shaded to support the orchid now even if it had survived the botanists. In some cases it might have been just bad luck: growing in the wrong place at the wrong time. Though it might seem heartless to say so, extinction's tally of the British flora is trivial.

What is far from trivial – but typical of countries all around the world, especially islands – is the much larger number of wild plants that are *nearly* extinct in Britain. Some 35 are considered to be critically endangered, that is, on their last legs, while another 90 are endangered, and no fewer than 220 are considered vulnerable, vulnerable to extinction, that is. A further 98 are slated as nearly threatened, that is, fast heading the same way. That makes 443 species in trouble, a full third of our native flora. Nearly all these plants are found elsewhere in Europe, in some cases commonly, and so there is little danger of them dying out entirely. The only endangered endemics – species confined to Britain and/or Ireland – are a quartet of hawk-weeds and a couple of whitebeams, all of which probably evolved locally and never spread very far. But these are not species in the full, sexual sense but 'micro-species' which reproduce without sex and thus don't evolve any further, and don't produce genetically mixed populations. Micro-species, it seems reasonable to suppose, do not last as long as full species. Their lifetime might be measured in hundreds, not hundreds of thousands, of years.

The obvious question to ask (but which, in my experience, is rarely asked) is why so many British plants are almost extinct, or threatened by extinction, or vulnerable to extinction, but are not actually extinct. They remain among the living, though only by a whisker in some cases. It suggests that our flora – which, after all, has survived in a densely populated farmed landscape for hundreds of years – is pretty resilient. Even today, when the demands on land are greater than ever before, there are usually still a few corners in which a scarce flower can find safety. Tall thrift, for example, has found an appropriate last resting place in a churchyard. Corn cleavers has survived against the trend in the experimental agricultural plots of Rothamsted Research, one of the oldest agricultural research facilities in the world. The New Forest, the wildest land in lowland England, has long been a refuge for wild flowers like small fleabane and pennyroyal, which would have been much happier in the pre-industrial landscape.

Certain plants might well have died out had we not stepped in to help them. Once a species has reached a late stage in its decline, the conservation machinery starts to whirr. Funds become available. Willing hands are found to devise plans, to buy or lease land for nature reserves, or to cultivate plants for a 'reintroduction'. Of those species considered to be critically endangered, strapwort has been cultivated in pots and trowelled into the gravel at its only remaining site. The lab at Kew has forged a technique for mass-producing lady's slippers. The last five plants of wood calamint were saved by micro-gardening, clipping over-shading boughs and weeding out invading nettles. Quite often it works, at least in the short-term. Although the lady's slipper may never spread far from the rocks in which it has been thoughtfully planted, the perennial knawel, a much humbler plant, multiplied vigorously once it was given a genetic shove by adding some nursery-bred plants. But intervention isn't always successful. For several species on the critical list there may be no convincingly wild plants left: fringed gentian, triangular clubrush, darnel, upright or city goosefoot. Perhaps the only reason why they have not yet been declared 'extinct in the wild' is that extinction sounds like failure. Conservationists do not like that word. And, like the rediscovered oil-beetles, there is always the hope that we may have missed a few.

Given that so many British plants seem to teeter close to the brink, what has caused their decline? Between them, our endangered plants are found in almost every natural habitat above the shoreline: open woodland, freshwater, fen, limestone grassland, well-grazed commons, mountain cliffs, sea cliffs, saltmarsh. Some of the endangered ones have pretty flowers but to notice others you need to peer closely at the ground, probably on your hands and knees. The most obvious losers are those which have effectively lost their habitat, like the corn buttercup, an inhabitant of weedy cornfields once well-known enough to acquire nicknames, such as devil's curry-comb, crow-claws or hellweed (it has spiky seeds). In other cases, it may be a biological failing, an inability to produce enough fertile seed, say, or hybridisation with a commoner neighbour.

An answer to the riddle of why some plants are threatened more than others was attempted by Kevin Walker, head of science for the Botanical Society of Britain and Ireland (BSBI), and colleagues, in *Threatened Plants in Great Britain and Ireland* (2017). They pinpointed certain key weaknesses that work against survival in the modern landscape. Among them are a short life (a lot of their listed plants are annuals), a limited ability to cast seeds far and wide, or with seeds that soon rot and so deny a plant one of its life insurance policies: seed-banks in the soil. They noted a plant that has nearly all these defects. The field gentian, *Gentianella campestris*, is a widespread but fast declining plant of grazed natural grasslands. It is short-lived, regenerates from seed only, lacks a seed bank and needs short, open turf to flourish. On top of that it is dependent on a supply of similarly declining bumblebees. If it was a schoolchild and we were writing its report, we would probably say 'must try harder'.

Another impediment to survival in modern conditions is an inability to withstand shade or competition from more aggressive plants such as nettles or coarse grass. It is striking how many of our rare flowers are not found in pristine habitats so much as human-made ones like hedgerows, village greens or even tracks. In the days when sandy commons were grazed by cattle, ponies and geese, or when the downs were kept short and open by the shepherd's flocks, these plants would have thrived. They found a living space in micro-habitats within habitats, in hoof-prints or anthills or some other temporarily open ground. A flower doesn't care whether bare ground is created by a bullock or a mammoth (or an ant), so long as the soil is not chemically altered or churned up.

It is the same in woodland. Few, if any, British woods are wildernesses. Most lowland woods have long been busy places. Until the twentieth century the main product from parish woods was small-bore timber cut on a coppice rotation. This, and the necessary lack of tall, over-shading trees, kept the woodland floor open and warm and also created a fair amount of surface disturbance. Plants had room to establish and spread while their sun-loving insect pollinators were equally happy. But machine-age disturbance of the kind caused by hydraulic diggers, by huge tractors and heavy trailers, is of a different order, less like scraping and more like ploughing. It doesn't disturb as much as obliterate. Meanwhile most woods have grown too shady to sustain their full potential of life.

Hence many of Britain's plants have long been adapted to a well-worked landscape. But once-traditional forms of husbandry have now been replaced by the efficiencies of the machine age: chainsaws, combine harvesters and chemical farming, and many once common species have been squeezed to the margins. Increasingly even those margins are looking precarious. BSBI's head of science Kevin Walker estimated that nearly every English county now loses at least one native flowering plant every couple of years, or in some cases every year – a local extinction rate twenty times faster than the national one. If maintained at the present tempo, logic would dictate that this slow slippage will result in no wild plants at all by the end of the present millennium. We barely notice these losses because we have short memories, comparing this year with the last, but not the one ten years ago, still less twenty years or fifty. We only know that loss is happening because plants are moni- tored by local flora groups, the patient work of hundreds of volunteers, out there searching and recording, often on their knees, come rain or shine.

In recent years, conservation has been successful at preventing further extinctions in the native British flora. The last probably native wild flower to die out was the swine's succory in 1980 (the stinking hawk's-beard might have died out in the 1990s but the situation is confused by reintroductions). You could say with some truth that in Britain native plants are no longer *allowed* to die out. But our rarest plants have come to be what Elizabeth Kolbert calls 'conservation-reliant'. We step in with ideas to compensate for the natural consequences of habitat loss. If a plant has lost its living space, we, in our orderly, method- ical way, provide new ones. Yet we can plan only for the known, not the unknown. Plant science is as imperfect as any other science, and climate change has the potential to scupper the best laid plans (just as the huge and unforeseen increase in deer scuppered the plans of woodland nature reserves). In fact climate change has the potential to shove every certainty back into the pot and give it a good shake. Change is the devil factor in conservation. It chips away at the knowns and swells the unknowns. Wherever we are going in the world we are going to have to live with that.

Extinction essay: the lost earwigs

What a lovely word. Earwig. Or, as it was in Old English, *eare-wicga*, 'ear wiggler', one of the oldest insect words in the English language. Appropriately earwigs are extremely ancient insects. There were already ancestral earwigs two hundred million years ago, scavenging, nibbling and creeping, oblivious of the parade of giant reptiles towering above them. They probably did rather well out of the meteoritic cataclysm that destroyed the dinosaurs. Earwigs, anybody might think, are indestructible.

The earwig shape is a tried-and-tested formula of long, thin, flexible bodies for slipping and squeezing into confined places – with that handy pair of nippers at the back in case of trouble. Like most insects, earwigs prefer the warmer, moistier parts of the world. There are only four species in Britain, one common, three rare (though experts on earwigs are even rarer). Time was when there was a fifth earwig, the granddaddy of them all, a tremendous bug variously called the tawny or striped earwig, but best known as the giant earwig from its impressive size. It was at least twice the size of an ordinary earwig and bore a massive pair of pincers resembling needle-nosed pliers. When threatened or annoyed, it raised these pincers up and over its back like a scorpion: a pair of sharp-ended nippers threatening to jab you in the face and then rip your nose off. Its scientific name is *Labidura riparia*, meaning, roughly, 'tail-pincers of the shore'.

In Britain, the giant earwig lived under dry seaweed just above the high tide mark, hiding during the day inside its personal burrow before emerging in the evening, cautious but hungry. Like other earwigs, it is a wonderfully protective mother, sitting on her eggs like a hen, and creeping out at night to find morsels along the beach to feed her babies. The male, meanwhile, is noted for having two penises, one, generally the right-hand one, for use, and the other presumably as a spare. The giant earwig was a reluctant flier, perhaps because the process of unfolding their delicate wings from an absurdly small case cannot be hurried, and then folding them back again, crease by crease, takes even longer, during which time it might easily be nabbed by a passing bird or spider. The giant earwig probably reserves flight for dispersal and spends most of its life on (or under) terra firma.

If our beaches swarmed with giant earwigs it would surely be among the best-known British insects. One could imagine kids daring one another to pick one up, or perhaps slipping a few into a sandcastle and watching them dig their way out. But few people ever saw one in Britain and today there may be none to see. Giant earwigs are virtually cosmopolitan in warmer parts of the world, but in Britain it always clung to warm pockets in the southern shore, mainly in Dorset and Kent. Unfortunately its favourite beaches have been reclaimed for building or improved for bathing – 'de-wilded', you might say. The first British specimen was recorded in 1808, at about the time people first started going on seaside holidays, while the last was in the 1940s. More recently, a sketch was made of an impressively sized earwig found somewhere between Sidmouth and Beer in South Devon. Unfortunately, the person then lost the sketch. So it may still survive. It is hard to be certain that something as small and unobtrusive as an earwig, even a 'giant' like this one, has gone for good.

Yet the giant earwig is not the largest earwig in the world. The winner is – or at least was – the great earwig of St Helena, that lonely rock in the South Atlantic, remembered chiefly for being Napoleon's place of exile. Tiny, isolated St Helena is/was full of endemic species found nowhere else: 50 land plants, 10 shore fishes and no fewer than 256 beetles. Some of them are now extinct. There is a rule of island evolution that big animals tend to get smaller and small ones bigger. Crete had a pigmy hippo but a giant swan. Corsica had an outsized shrew, Majorca a mega-dormouse, Timor a humungous rat. Even Britain, not long isolated as an island, has a slightly bigger vole living in Orkney and a slightly meatier field mouse on St Kilda.

By the same rule, St Helena had a giant earwig named after Hercules: *Labidura herculeana*. It was the size of your forefinger and was armed with a truly formidable pair of nippers. It lived in tunnels among the rocks on the island shore. It could afford to be a giant of giants because, until humankind first set foot there, the island had no resident rats or mice, indeed no predators of the sort that might be expected to make short work of *Labidura herculeana*. But unfortunately rats and mice did arrive with the visiting ships, and they probably did eat the earwigs. Except that, as things turned out, humankind beat them to it by trashing the earwig's habitat. The rocks where it burrowed and crawled were needed to extend the airstrip, to build more houses, and to improve the harbour. (Some also blamed entomological expeditions in the 1960s which evidently collected heavily.) The last time anyone saw a live Herculean earwig was in 1967. Several unsuccessful searches later *Labidura herculeana* was formally declared extinct. They say that, by careful searching, you can still find dried up fragments of earwig along that lonely shore, especially those indestructible, amber-like nippers.

Left alone, earwigs are among life's great survivors, like cockroaches. They can eat almost any vegetable matter as well as small insects and spiders (they have a penchant for insect eggs too). They hide away snug inside crevices and cracks. They can look after themselves with their nippers and they also look after their young, packing away several broods within their short insect lives. When I was a boy you found earwigs everywhere, in the garden, under flowerpots, in the compost heap or tucked inside bitten blooms of dahlias and chrysanthemums. You met them indoors too, after they had crept in under the door or through a crack in the kitchen window and made a new home behind the wallpaper. My mother told me she would never use dahlias as table decorations in case it introduced a nest of earwigs to the house. Perhaps back in the days when we slept on straw in huts, we might even have shared a bed with them. Those people might or might not have been able to endorse the old tale of how earwigs will eagerly crawl into the warm and moist shelter of your ear.

But are earwigs, like so many other insects, creeping out of our lives? On BBC's *Gardeners' Question Time*, listeners sometimes ask the experts what has happened to the earwig: where have all the earwigs gone? They are still out there, to be sure, but in vastly reduced numbers. They no longer seem to lurk under every rock or in the food left out for the hedgehog. No one knows why, though we might sense, vaguely, that our environment has become too tidied up or too polluted by chemicals. Many of today's children have probably never even encountered an earwig. Unlike Britain's giant earwig and St Helena's Herculean earwig, it is unlikely to die out completely – that would take a real apocalypse. But, like moths on the window-pane or bumblebees in a meadow, they are less familiar than they used to be. Today's children will not come across earwigs so often (or know them when they do), nor ladybirds or river mussels or stag beetles. It represents a kind of cultural extinction, this piece-by-piece excision of the familiar from our lives. When once common animals start to disappear it gets scary. You wonder what has happened to them, and what it might mean. You begin to see what the biologist E. O. Wilson meant by the Eremocene, the New Age of Mankind, all alone in the world except for his pets, his herds and flocks, his useful animals.

Chapter 4

Extinction – and how to avoid it

The trouble ain't that there is too many fools, but that
the lightning ain't distributed right.
Mark Twain (attributed). Merle Johnson (1927),
More Maxims of Mark.

All species become extinct in the end. For some it will
come sooner; others later, perhaps much later (some plant
and animal forms have changed little over tens of millions
of years). In the eternal contest of life, many things help
you to avoid early extinction. A broad natural range and
an ability to adapt to changing circumstances are important
assets, but in the modern, human-dominated era the greatest
gift of all is the ability to take advantage of whatever
humankind happens to offer, whatever crumbs he might
let fall from his lordly table. The world's most successful
species are now those which associate with us.

Another useful survival trick is to reproduce fast and
often, building up a vast swarm of siblings. Nor will our
successful species be over-fussy when it comes to habitat
and home; it will make do on our wastelands and on
whatever scraps of natural habitat remain. It won't be a
fussy eater either because species that anchor themselves to

a limited food supply may face starvation in times of change. Above all, our species won't be a habitat specialist. It worked in the past but it won't work now for natural habitats are in retreat the world over. Our species will also need healthy genes to survive. Inbred species are vulnerable; their immune defences get breached, they catch diseases and, as we know only too well, from our own current experiences, disease can quickly turn lethal and global. Oh, and it's not a good idea to be large and beautiful either, or people may want you as a trophy.

Here I offer a dozen precepts of advice to our imaginary species seeking to enjoy a long life on earth. Starting with perhaps the most important of all . . .

Don't live on an island

About 200 species of birds have gone extinct since 1500, that is, four per decade or one for every fifty species alive today. That is much faster than the purely natural rate, which is more like one species in every 700 years, or just one unlucky bird since the Middle Ages. But even four per decade is quite a stately pace of lost birds compared with the rush of what is happening today. Some 1,200 more species are heading towards the cliff, and that is one bird in nine. These losing birds are similar in kind to the already lost ones. There have been particularly steep losses among the pigeons (so in their case the meek did not inherit the earth) and among parrots, with a smaller but still significant count among the world's owls, raptors and vultures, curlews, ducks, starlings, kingfishers and woodpeckers. Worst hit of all are the rails, those ground-living birds of land and water, many of which lived on isolated islands, while some are, or were, also flightless, a deadly combination of frailties. If you want a quiet and trouble-free life, don't be an island rail.

In fact, don't be an island bird at all. I have in front of me a list of the 80 best-known recently extinct birds and around 80 per cent of them lived on islands, often just one island. That chary existence is reflected in their names: Laysan duck, Lord Howe Island white-eye, Jamaica least pauraque, Grand Cayman thrush. The extinct bird capital of the world is Hawaii, once an abode of unique and often brightly coloured birds that fed on nectar and helped pollinate the island archipelago's flowers, hence their names, honeyeaters and honeycreepers. Of 142 Hawaiian birds found nowhere else on earth, 95 have gone already and the outlook for the remainder, for the poor *akikiki*, the *i'iwi* or the *kiwikiu*, looks far from rosy. It is not far from the truth to say that pretty well every native bird on Hawaii is threatened. Of course, there are lots of other birds there that are not threatened at all, but those are the 'introduced' ones. The ones that helped to exterminate the natives.★

Next after Hawaii is the island of Guam, which has lost 60 per cent of its native birds, mainly from an accidentally introduced egg-eating tree snake. Also high on the loss list are the Mascarenes, including Mauritius, and the tiny islands of the Tasman Sea between Australia and New Zealand. But name pretty well any ocean island in the world and the chances are there will be a lost bird or two. New

---

★ Some of the honeycreepers may have died out from contracting avian malaria from an introduced mosquito. Today the surviving species of honeycreeper live on the wooded flanks of mountains too high for the mosquito to reach – yet.

Caledonia, for instance, has lost its charming little parrot, the Caledonian lorikeet. The Society Islands have lost a reed warbler, oh, and a thrush, Wake Island lost a rail, the Solomons a pigeon.

What went wrong for all these birds? One answer is low numbers and lack of space, a rapidly dwindling space once their islands were settled by humankind. Many were forest species that simply ran out of forest, or at least the right kind of forest. In some cases the local birds vanished so quickly, in a mere blink of the historical eye, that they were only discovered after they were dead, from bones, feathers, eggshells or whatever other poor remains they left behind. But the real killer was (and is) invasive species, the cavalcade of livestock, rats, cats, snakes, even ravenous snails, which accompanied us on our voyages and settled wherever we did, and made a new home for themselves at the expense of the native fauna. Isolated island birds had no experience of predators like cats and rats, for they never needed to deal with anything faster-moving than a land crab. Evolution had provided them neither with the wits nor the behaviour to protect their eggs or chicks or avoid being eaten themselves. And so they vanished down the gullets of animals, including *Homo sapiens*, which, in evolutionary terms, had no right to be there.

So don't be an island bird. In fact, don't be an island anything. Hawaii might be the lost bird capital of the world, but Madagascar holds the record for extinct and endangered reptiles, and the many islands of Indonesia are full of threatened mammals. On Hawaii even the native moths aren't safe. If you have to live on an island make sure you are a seabird with powerful wings to take you away when things turn sour. Which they will, and sooner than you think.

Don't depend on snow

As the world's climate warms up, animals that depend on snow and ice are being squeezed. Between 1979 and 2016, the Arctic lost 43 per cent by area of its ice, and what was left grew thinner. In total, more than three-quarters (77 per cent) of the Arctic sea ice has melted away in our lifetimes. Soon the top of the world will no longer be white. Nor does the snow linger so long in mountains, and all of this has serious implications for cold-adapted wildlife, above all animals which turn white in winter, such as the European ptarmigan and mountain hare, or the North American snowshoe hare. The pure white fur or feathers that camouflage them so well in snow makes them almost comically conspicuous against a dark background. And that, of course, makes life so much easier for their predators (but if the ptarmigan and hares are killed *en masse*, then the foxes and lynx and eagles will lose their prey: in the end everyone loses).

The most famous victim of Arctic melting is of course the polar bear. Living where it does, in the cold, remote roof of the world, it is hard to estimate the numbers of the world's biggest terrestrial carnivore (all right, it might tie for first place with the brown bear of Kodiak island), but there are believed to be between 22,000 and 31,000 animals in total.

Unfortunately for it, the polar bear is a maritime bear that relies on sea ice for hunting its main prey, seals. Because Arctic sea ice is decreasing and thinning, the bear now has to roam further to find food and so uses up more energy than before, especially when it is forced to swim from flow to flow as the ice breaks up. On top of that, the winter ice is now melting before the bears have built up sufficient reserves of blubber to tide them over when food is scarce. And on top of *that*, their blubber concentrates pollutants like PCBs passed on by the seals. Resourceful bears may get by for a while on berries or fish or foraging human rubbish, but moving away from the ice brings them into contact with humans – never a good idea – and also with well-adapted brown bears in a contest the polar bear is unlikely to win.

All this is making life hard for polar bears. As recently as 1986, the animal was regarded as 'not at risk' in North America. A few years later, its status was uplisted to 'special concern', and when its numbers were reassessed in 2015 that was further promoted to 'vulnerable'. At the next assessment, in 2025, the polar bear will almost certainly become officially 'endangered'. Scientists predict that two-thirds of the polar bears in North America will have gone by 2050. The global population is expected to decrease by a third within the same period. Hence this wonderful bear, with its snowy charisma, has become the pin-up species for climate-change victimhood.

Warmer winters also spell trouble for Arctic or alpine flowers that rely on a covering of snow to protect their tender shoots from the frost and winds. Mountain plants include some of the world's most attractive wild flowers: saxifrages, primulas, lilies, gentians, rhododendrons, edelweiss. Many of them have a limited range, isolated as they are in their mountain domains. Warming makes them vulnerable to competition from more aggressive plants. A very detailed study of Ben Lawers, the most flora-rich mountain in Britain, found that lowland plants were edging uphill, and that the natural mountain vegetation was not only shrinking but becoming less rich, more homogenous. Late patches of snow also form distinctive micro-habitats whose insects, emerging in swarms as the snow melts, offer an important food source for upland birds when they are rearing their chicks. But such snow beds are becoming less frequent and melting earlier. Ben Nevis, or 'the ben of snow', Britain's highest mountain, may soon be completely snowless by summer. In the Cairngorms, Britain's only sub-Arctic plateau, snow used to linger well into the summer, but in most years now it has gone by June, even in the once well-known snow patches like Ciste Mhearad ('Margaret's Coffin'), where I practised ice-axe braking in colder, better days. At least two Arctic lichens that were confined to such places seem to have disappeared altogether.

Growing where they do, mountain plants are hard to monitor year on year. However if predictive models based on climate data are right, the Cairngorms will become almost snowless even in winter by 2080. That would be bad news for plants and animals of the high tops, and bad news too for the fishermen below. The salmon, for which they pay top rates, depend on cooling torrents of snowmelt to spawn in Highland rivers. Unless something can be done about climate change, it may well be that mountains, not just in Britain but worldwide, will become another extinction hotspot.

Don't over-specialise

In this ever-changing world it helps to be adaptable. Specialisation may make you supremely suitable for a particular set of circumstances, but once those circumstances change you are sunk. That might be what propelled the large blue butterfly to extinction in Britain. Its isolated populations dwindled away one by one until there was only a single small colony left, and that wasn't enough to save it from two cool, wet summers in a row. The butterfly was declared nationally extinct in 1980. A larger, healthier population might have made it through, but small ones are vulnerable to genetic drift and chance events. Without a vigorous nearby colony to replenish it when the good times return, it's over.

The large blue was later 'reintroduced' (the misleading jargon word for foreign introductions) to England from Sweden. By one of life's little ironies, Jeremy Thomas, a pioneer of the study of butterfly ecology, had just worked out how to save it when it died out. The large blue has a peculiar and complex life cycle. For the first few weeks of its life its caterpillar lives just like any other caterpillar, quietly munching the leaves of a plant, in this case wild thyme. Then, at a certain stage in its life, it loses interest in plants, is picked up by a passing ant and taken down into the ant nest. There, while offering the ants a sweet drink from its glands, like a cow, the caterpillar turns carnivore and helps itself to the ant's own grubs. Eventually, and still in the ant's lair, the now plump caterpillar turns into a chrysalis, and then into a butterfly, crawling out into the sunlight with guardian ants still clinging to it.

All this was known before, but what Thomas discovered was that the tricked victim wasn't just any old ant but a particular one, a red ant called *Myrmica sabuleti*, and that too had specialist needs, namely warm, south-facing hillsides with short, fine grass. In Britain, only nature reserves could provide what both ant and butterfly required, and, since no one then understood what was going on, they didn't.

From the nineties onwards, similar-looking large blue butterflies from Sweden were introduced to sites that had been carefully managed to achieve the necessary short turf and copious supply of ants. The Swedish butterfly was judged to be genetically near identical to the native English form; or close enough anyway. It took a lot of work and dedication to bring the project to fruition, and much of the monitoring and site maintenance was done by volunteers. They say the butterfly is now sufficiently well-established to sustain its numbers and even spread into new places. All the same, it is still a Swedish immigrant living in England. Whatever distinctiveness the native form might have evolved through 8,000 years of isolation has gone for good. The large blue's presence in England today comes from decisions taken by well-meaning and highly motivated humans, not natural selection.

The story of the large blue illustrates the perils of over-specialisation, especially in times that call for maximum adaptability. British butterflies, and maybe butterflies generally, divide into habitat specialists, rooted to their patch by their needs, and the 'wider countryside' ones which move around, visit gardens and make the best use of whatever opportunities they find. The first group are mostly declining, the second mostly hanging in. For a truly suicidal degree of specialisation we could turn to a moth called the viper's bugloss (named after a plant they thought it fed on). For reasons best known to its genes, this moth chose for its sole larval food plant the rare and insubstantial Spanish catchfly, and not the whole plant either but just its tiny seed capsules. And so each caterpillar needed a great big patch of catchfly to feed up and complete its cycle, and, in Britain, such patches of Spanish catchfly are no longer to be found. And so the moth shrugged and died. As Kenneth Horne used to say on BBC Radio's *Round the Horne*, 'there's a lesson for us all there'.

Coral is, in a way, a specialist. Coral reefs form in clear, shallow water of just the right temperature, neither too warm nor too cool. They need plenty of overhead sunshine because coral is the animal equivalent of lichen: the little coral polyp gains most of its nutrition and energy from the photosynthesis of microscopic organisms embedded in its tissues. It is they that give the reef its colours, and they that produce the needful sugar which they kindly pass on to the coral animal. Unfortunately warming and pollution cause the micro-organisms to malfunction, at which the coral indignantly expels them from its midst. This in turn means the coral loses its colour and 'bleaches'. Coral can survive for a while in its bleached state, and even put forth strange colours of its own, but ultimately without its partner it will starve and die, turning the reef into what one commentator called 'an ossuary of rubble'. And that's the trouble. Climate change is warming the sea, and coral reefs are bleaching and dying all round the world. It took two million years for the Great Barrier Reef to form. It may take less than a century to die. And without coral what kind of future will remain for the 4,000-odd species of fish that depend on it, not to mention the uncountable species of equally dependent small life?

Don't be enormous

The interval between the discovery of Steller's sea cow and its subsequent extinction was very short: just twenty-eight years. So short, in fact, that only one man ever studied this huge animal and that was Georg Steller who brought this great lost animal to life.

A German naturalist and physician aboard a Russian ship with a Danish captain, Steller wrote about the strange animals he saw on his Arctic voyages in *De Bestiis Marinis* (The Beasts of the Sea), published posthumously in 1751. He first sighted the great sea cow as the ship's physician on *Vitus Bering*'s disastrous second voyage. A giant relative of the manatee and a very distant relative of the elephant, the sea cows were spotted wallowing in the kelp beds surrounding the Commander Islands in what is now the Bering Sea. As Steller discovered, once they managed to hook one and haul it on-board, this beast was enormous, thirty feet long and weighing, Steller estimated, about twelve tons, more than a bull elephant and as much as a medium-sized whale (though modern estimates are slightly lower, in the eight to eleven ton area). Its intestinal tract measured 150 yards. Whales aside, Steller's sea cow was the largest animal in the world. It spent its whole life afloat in shallow water, unable to dive very deep, feeding on kelp. According to Steller, the herds made sighing and snorting sounds, as if distressed at being discovered – and, if so, they were right to feel that way. Steller's sea cow was a rare survivor of the Ice Age megafauna, but not for long.

It used to be thought that the sea cow was simply hunted to extinction. It was a useful animal, especially to starving, scurvy-ridden sailors enduring a long Arctic voyage. It was also a very slow and defenceless one. Its meat was rich and beefy. Its blubber, melted into fat which Bering's men drank by the mug-full, tasted of almond oil. They could even make butter from the sea cow's rich milk, and replacement boots from its thick, waterproof hide. The animal assisted in its own destruction by gathering round the one the sailors had hooked, and so putting themselves in danger. All the same, exterminating every last one of them in tiny boats in icy, tossing seas, and in so short a time, does seem quite a feat.

More likely, scientists think now, it had something more to do with another animal target: the sea otter, and this is why. In the days when sea cows were still around, the sea otter was hunted relentlessly for its fur. The price of otter fur compensated hunters for the arduous and dangerous journeys along the Arctic shores of Russia and America. The otters fed chiefly on sea urchins, which in turn fed on the kelp. By controlling the numbers of urchins, they were the 'keepers of the kelp'. But once otters were nearly exterminated, the urchins lost their predator and so multiplied into vast numbers. With less kelp to feed on, the sea cow herds were in trouble, for they needed a large and constant supply of it. The animal was probably already at a low ebb, having been hunted by local people on the northern shores of the Pacific rim for centuries. By the time Steller discovered it, there may have been only 1,500 or so left. By destroying the delicate balance which tied sea cows, otters and kelp together in mutual harmony, extinction was inevitable. But even so, the speed with which it happened is unsettling. The message seems to be clear: don't be large, don't be slow and don't hang about when the ship draws near. But above all, don't put your trust in kelp. Or, indeed, in any other plant liable to vanish when nature is thrown out of kilter.

Don't be a shark

These days if you are a shark, your chances of living a long and fulfilling life are slim. Especially if you are large and live in shallow water. All too swiftly you may find yourself among the 100 to 200 million sharks caught by hooks or fishing nets every year, and that is about 1.5 million tonnes of shark (these are guesses at an unknowable figure, but a recent estimate put it very broadly at between 63 and 273 million dead sharks in the year 2000 alone). Some of the dead sharks are sold as meat, but many are killed primarily for their fins in order to make that tasteless but expensive delicacy, shark fin soup. Much of the holocaust of the world's sharks is unintentional, with the sharks scooped up incidentally as a bycatch. More than half of these unwanted sharks may be chucked back overboard, more dead than alive, and probably shortly to be dead (and definitely so if they no longer have fins). The annual catch rate, which rose steadily from 1980 to 2000 but has declined somewhat since then, represents between six and eight per cent of the global stock of sharks. This may not sound much but it is unsustainable; the falling numbers of sharks proves that. Unlike bony fish, many sharks have a low fecundity and so also a low replacement rate, at about 30 per cent lower than the average catch rate. Sharks tend to be long-lived and slow to mature (the Greenland shark matures at 150 years old!), with long pregnancies and relatively few babies.

What effect is it having? Shark experts around the world meet regularly to thrash out the statistics, which, they tell us, are 'chilling'. The 2020 Global Shark Trends study, run by the IUCN, indicated a 71 per cent decline in pelagic sharks worldwide since 1970: in other words nearly three-quarters of the large sharks that swim in the surface waters of the world's oceans have disappeared in the past fifty years! The year-on-year figures suggest that a shark was eighteen times more likely to come to a soupy end in 2000 than in 1970. Given all that, it is hardly surprising that three-quarters of the thirty-one most prominent shark species assessed by the study are believed to be in danger of extinction.

The same dire statistics apply to the shark's distant relatives, the rays. The numbers of the giant manta ray, for instance, have fallen by between 50 and 80 per cent over the past three-quarters of a century. Although a very popular fish with divers – huge, impressive and harmless – the manta has been targeted by Asian fishing fleets for its gill plates which are used in traditional medicines. Of the 600-odd species of ray, about 200 are threatened. All species of giant guitarfishes (named after their bizarre shape) and all but one of the wedgefishes (shark-like rays with particularly valuable fins) are deemed critically endangered. For sharks the equivalent figure is 153 species out of 535 (but many species are 'data deficient', meaning we don't know what is going on).

How many shark and ray species are actually extinct? None for certain, though at least three, including the well-named lost shark, have not been seen for decades. In the vast spaces of the world's oceans there has to be a long absence before there can be any certainty on that score. Many others may have vanished before they could be collected and described by scientists. What we do know is that in some areas, such as the grievously over-fished north-east Atlantic, fishing has virtually wiped out some species. The common skate, for instance, needs renaming: the 'extremely rare skate' perhaps. A small shark which older people may remember eating with chips as 'rock salmon' (a name now outlawed by consumer legislation), is the spiny dogfish, also known as the spurdog or piked dogfish from a venomous spine attached to the dorsal fin. Perhaps we also remember dissecting them in the school biology class – those rough, stiff bodies reeking of formaldehyde. The spiny dogfish was formerly one of the commonest predators in the sea and supplied a huge fishery. But by the time all landings were prohibited along the Atlantic shore of Europe ten years ago, only 10 per cent or so remained from its original numbers, and it is now assessed as critically endangered. The spiny dogfish had been driven to near extinction in the north-east Atlantic by its one weakness. The female does not reach sexual maturity until towards the end of its long life, at twenty-five to thirty-five years. And it also has one of the longest pregnancies in the animal world, taking up to two years to give birth to its pups. The result: massive unsustainability.

So don't be a dogfish and don't be a shark. Or a ray. Safety for your species will require worldwide agreement on fishing quotas and their strict enforcement by an international navy of armed and incorruptible patrol boats. Will that happen? All one can say is: it hasn't happened yet.

Don't be a frog either

I will never forget the first time I saw a poison frog. We had gone for a swim in the forest in Costa Rica, and there it was, squatting on a rock close to the waterfall, bright strawberry-red with small black spots, and about the size of a 50p coin. It was there and then it was gone. It didn't hop like a European frog but seemed simply to vanish, leaving behind a sense of vivid colour, a spot of red behind the eyelids. Later I also saw several of the little mint-green and black ones, seemingly freshly painted, and looking more like ornaments than living animals. They sat motionless, secure in their venom-laden skin.

I never did see the most famous of them all, the golden frog, because by the time of my wildlife holiday it was already extinct in the wild, or nearly so. The golden frog was one of the largest and most toxic of all the poison frogs. There was sufficient venom in its little body to stop a man's heart. They say it was also the most intelligent frog, an awareness hinted at in its beady black eyes. In captivity, golden frogs learn to recognise their master, the one that brings it food. In Panama, it was thought to bring good luck, so much so that they even put its portrait on lottery tickets. When a golden frog died it was supposed to turn into an ingot the size of a shiny gold coin. Scientists investigating its sudden disappearance in the 1990s found plenty of dead frogs but unfortunately no golden coins, just shrivelled carcasses, often turned upside down with legs akimbo. If they ever did bring good luck to lottery ticketholders, it certainly didn't extend to the frogs themselves.

The message is that being a frog, especially in the tropics, is no longer a wise choice of lifestyle. You are probably going to die soon from a very unpleasant disease. What killed the golden frogs, and is killing other frogs all round the world, and by the million, is a fungus that proliferates on their skins which, in effect, chokes them to death. Frogs breathe partly through their skin, but this necessarily thin, permeable membrane also means that they will readily absorb any nasty human-made chemicals, such as pesticides, coming their way. For the same reason they are vulnerable to infection by this skin fungus to which they seem utterly defenceless. Mankind has probably been the unwitting vector of this disease through the global trade in African clawed toads (*Xenopus*) used for pregnancy tests: the poor toad comes out in spots when injected under the skin with a pregnant woman's urine. The water-born fungus spreads along the myriad streams of the mountain forests, killing individual frogs within days, and, it seems, the entire world population of golden frogs inside just a few years. With the golden frog went its compatriot, the golden toad. Almost the last ones ever seen alive in their natural habitat were filmed, in 2006, for Sir David Attenborough's BBC series *Life in Cold Blood*.

The death of golden frogs and toads sparked much activity in conservation circles, with experts from around the world gathering to compare notes. Their stories were essentially all the same: more disease and fewer frogs. By 2004, they reckoned that of 8,000 species of amphibians in the world, most of which are frogs, nearly 500 species were in danger, most of them critically so. It was impossible to say how many were already extinct, but the estimate was at least nine species and perhaps as many as 122. Since then, the proportion of amphibians on the danger list has grown again from 32 to 42 per cent. (Locally things can be even worse; for instance two-thirds of India's frogs are considered to be endangered.) The extinction rate among frogs and their kind is calculated to be 211 times higher than the natural rate, a slippery helter-skelter of fallen frogs. Compared with that, the depletion of poison frogs through the pet trade, and also from the ongoing destruction of their habitats by logging and fires, albeit considerable, seem more like an extinction top-up.

So, don't be a frog. Or a toad.

Life has become suddenly uncomfortable for amphibians. They appear naked and exposed to the world's furies. Numbers are falling everywhere. For example, a survey of 153 ponds and ditches across Britain undertaken by the charity Froglife suggested that the common toad has declined by two-thirds over the past thirty years, that is, by 2.26 per cent per year. A way of life, which has served amphibians well for a hundred million years and produced a dazzling array of colourful and bizarre species, has suddenly hit the buffers in this over-crowded, polluted, diseased world we live in.

Be adaptable

We are what we are because of our genes. To be a successful species, to last at least a few million years, you require plenty of genetic variation. Diversity is the spice of life, the raw material of evolution. To put it another way, without sufficient genetic variation for natural selection to work on, we won't evolve. When bad times come, we will lack the ability to adapt, and then, as a species, we will die. We tend to think in terms of individual animals but evolution works on large numbers, that is, 'populations'. The death of an individual, however sad for its relatives, means nothing. What matters is the flow of genes within the mass: the genetic health of the species across its range.

Looking around, we can sometimes spot species with sound genetic health. Ourselves for example: our genes offer different skin colours, different eye colours and all kinds of other gene-controlled characteristics — tall, short, healthy, not so healthy. Lions seem to have it too, with long and short manes, different skin folds and slight differences to their skulls. Where such differences correspond with genetic markers, a species may have a lot of life left in it. Lions have been around for nearly two million years, and, if we leave them alone, which we won't, they have what it takes to survive two million more. Cheetahs, on the other hand, have very low genetic variation. In zoos, and in the wild too, they are afflicted by an often-fatal viral disease. Lions, subjected to the same virus, survive it.

At the more critical end of things, let's take the black-footed ferret of North America. Reduced to a tiny population by habitat reduction and disease, the ferret seemed doomed. To save them dying out altogether, the last known wild ferrets were rounded up for captive breeding, although only seven of those weakened individuals were able to give birth. Fortunately, tissue taken some time before from a healthier and much more genetically diverse individual called Willa had been preserved, frozen in a 'biobank'. In 2020, it was used to create the first 'test-tube' ferret, a clone of its long-dead parent. This little pup, christened Elizabeth Ann, is now the most genetically valuable black-footed ferret alive, the last and best hope for its species. Several generations on from Willa, there are now between 250 and 350 black-footed ferrets in captivity and perhaps 300 more in the wild. But without the genes preserved in that biobank the ferret may well have been extinct by now.

A similar lab-based technique was used to create the first 'test-tube' Przewalski's horse (pronounced 'shu-*val*-skees'), an endangered wild horse from Mongolia that had been declared extinct in the wild. Unlike North America's mustangs, it is a true wild horse and not an escape from domesticity (or at least, as its DNA seems to indicate, a now wild survivor from a very ancient line of domesticated horses). The animal would be extinct had it not been for captive-bred individuals in zoos. At one time, only twelve pure-bred examples existed. Interbreeding horses from different zoos helped the animal recover from this near-fatal genetic bottleneck and, by the 1990s, there were enough horses to release some back into the wild, on the steppes of Mongolia. Meanwhile, tissue from one 'genetically important stallion', preserved since 1980, has been used to produce a cloned foal, which is currently doing well under the care of its surrogate mother, a normal domestic horse.

Of course, using biotechnology to revive a species is expensive and, at present, is open only to animals that carry prestige – in other words species whose extinction would reflect badly on us. By contrast, what hope is there of reviving the fortunes of the little-known white-headed langur monkey of south-west China? Genetically it seems to have reached rock-bottom through inbreeding, and, in consequence has problems other than the usual one of a shrinking habitat. The monkey seems peculiarly prone to disease and to parasites. Its reproductive rate is disappointing, and in general the species seems to lack resilience. Plainly a failing species, the poor monkey's best hope lies in mixing its two isolated populations and so adding a little fresh blood to the sample. Even the white-headed langur is a picture of genetic health compared with the island San Nicolas fox which is a genetic flatline, a diversity zero. Scientists found no variation at all within the last 300 surviving foxes. They are all the same.

Like all English people of a certain age, I have direct

experience of what can happen when a species flatlines. Before the 1970s, much of the English rural landscape was characterised by elms, tall trees that towered above the hedgerows in billows of foliage, like arboreal clouds. As children we knew them well because English elms were impossible to climb. Unlike oaks and chestnuts, they had dense trunk-palisades of leafy twigs that were impossible to break through. But they were loved by poets and by birds, especially rooks. They seemed as eternal as the real clouds above.

And then one bad day in the early seventies, Dutch elm disease struck. The agent, a fungus carried by a beetle that burrows under the bark, was introduced to Britain in imported timber. Within just a few years, nearly all those tall elms were dying or dead, soon to be felled and erased from the landscape. Some survive in hedgerows, where new shoots grow as suckers from the still-living roots, but long before they can turn into trees, the beetle dives in again and reduces them to a dead, antler-like state. As a tall tree, the English elm is effectively extinct (a small population of still healthy trees exists in Brighton on the south coast).

Why did this happen? Although England had about 25 million elms, their genetic variability was negligible. Once thought to be a native tree, molecular and literary evidence suggest otherwise. The elm was probably introduced by the Romans because it was perfect for supporting and training grape vines. English elm produces pollen normally, but hardly ever manages to set fertile seed. Its spread was either purely vegetative, by means of root suckers, or assisted by planting. And so, when disease struck, the tree had no resistance. The elms were genetically all the same, and they all died in the same way. They are missed only by those of us that remember them, the same as many extinct species.

Be careful where you leave your eggs

I first came across the great auk on a plate in *Thorburn's Birds*, a surprise bestseller of the 1960s. The artist Archibald Thorburn had chosen to paint it squatting on a rock, looking up in apparent envy at a guillemot perched on a ledge above. Being unable to fly, the auk could not reach cliff ledges, and so was forced to lay its single mottled-brown egg at the base of the cliff, just out of reach of the breakers.

The extinction of the great auk is one of the best-documented in history. It was the largest and heaviest northern seabird, standing 80 centimetres tall and weighing about 5 kilograms. Though it could not fly, its stubby wings functioned effectively as flippers. Being black and white, the great auk looked a bit like a penguin – some say it was the original penguin – and that is what the sailors called them as they slaughtered the helpless birds for their flesh and their fat and their feathers. Before humankind, the great auk had few natural enemies. For its way of life, chasing down fish and nesting on rocky islands, it had no more need of flight than a penguin, but its flightlessness spelled disaster once the boatmen stepped ashore. The great auk was also unlucky. One of the last substantial colonies was sited at the base of an active island volcano off Iceland. When it erupted, in 1830, the surviving auks moved to another, smaller island called Eldey but they didn't last long. Unusually we have an eye-witness account of the fate of the very last bird, by the man who killed it:

'It walked like a man . . . but moved its feet quickly. [I]

caught it close to the edge, a precipice many fathoms deep. Its wings lay close to its sides – not hanging out. I took him by the neck and he flapped his wings. He made no cry. I strangled him' (Fuller 1999).

The stuffed remains of that bird are now on display in the Belgian Institute of Natural Sciences. Its eyes and internal organs are held in glass jars in the Zoological Museum of Copenhagen. Altogether some seventy-eight great auk skins and seventy-five eggs have survived, all from that single island and all catalogued and numbered (for his book, *The Great Auk*, Errol Fuller photographed practically every known egg). The £9,000 paid for a great auk in 1971 by the Icelandic Museum of Natural History entered the *Guinness Book of Records* as the most expensive stuffed bird in history.

The great auk's likely weakness was its egg-laying habits. Its closest living relative, the razorbill, tucks away its eggs in rock crevices on inaccessible cliffs, but the auk was restricted to the places it could waddle up to. And, unfortunately, any place that a walking, metre-tall bird could reach was equally accessible to a hungry islander or trophy hunter. Given that the great auk laid only a single egg on the bare rock, and the likely time needed for incubation and feeding the chick, it was unlikely to lay again once its egg had been stolen. On the isolated isles of St Kilda, off the Scottish west coast, great auk eggs were eagerly gathered, and that may be why the bird had deserted those parts by the end of the seventeenth century. By the nineteenth century, nowhere on earth was safe. Without legal protection properly enforced, there could be no hope for the great auk.

The great auk made a posthumous literary appearance

as the Gairfowl in Charles Kingsley's *The Water Babies* (1863). 'If only you had had wings,' says Tom, 'you might have flown away.' 'Soon I shall be gone,' the Gairfowl answers, weeping 'tears of pure oil', 'and nobody will miss me'. And then the unsentimental author quotes Tennyson: 'The old order changeth, giving place to the new', a pithy enough summation of what extinction means.

Incautious egg-laying may be at the heart of one of the mysteries of evolution: what happened to the ammonites? I have walked beneath the cliffs at Lyme Regis, on the Dorset coast, and spotted ammonites by the dozen embedded in the fallen rocks, some only the size of a penny, others up to the size of a car tyre, all of them tightly coiled, like a Catherine wheel. These are only the shells of the creature, or at least the casts of shells, and the soft-bodied animal inside them has not been preserved. The ammonite probably had tentacles, like the present-day nautilus, but what it fed on was, until recently, a mystery. Now close examination of the micro-fossils that once formed part of its jaws, suggest that some, if not all of them, fed on plankton, which might explain why they were so incredibly common in those ancient seas (back then there were no huge baleen whales to hoover up the plankton).

Other hints in the fossil record indicate that the ammonites, like their modern relative, laid soft, gelatinous eggs whose hatchlings swam in the plankton near the ocean surface. If so, it was a strategy that served them well – right up to the day a meteorite struck the earth and boiled the oceans. That was the end of the eggs and the end of the ammonites too, but the ancestors of the nautilus, presumably living much further down where the water remained cool, survived. Thus the story of the ammonite becomes a watery version of the fable of the grasshopper and the ant. The grasshopper has a free and easy life, until the winter arrives, and then it starves. The more prudent ant anticipates the future and survives. So it was, or might have been, with the 'frivolous' ammonite and the 'cautious' nautilus. All the same, the ammonites had a good run, and they would probably have been gobbled up by modern sharks in any case.

Leave yourself plenty of room

Back in the 1950s, my old colleague Norman Moore spent a lot of his spare time recording species on the heaths of Purbeck in Dorset. When Thomas Hardy based his novel, *The Return of the Native* (1878), on those open vistas of heather and bog, they stretched in an almost continuous line from Dorchester to the New Forest, thirty miles of marginal land full of wildlife. By the 1950s, however, Hardy's unified heath was fragmented, now more of a patchwork of over a hundred mini heaths of varying size, interspersed by farmland and housing. Less than 20 per cent of the original heath survived although that still makes it one of the largest areas of lowland heath left in north-west Europe.

What Moore discovered is that the larger fragments of heath contained more species than the smaller ones, and especially those most characteristic of heathland such as sand lizard, smooth snake and the silver-studded blue butterfly. The smallest fragments had lost nearly all their special plants, animals and insects. They had effectively become islands, isolated from the rest, and were now bleeding species year on year. Moore's project bore out what has since become one of the dictums of conservation, that the biggest areas contain the most species. This holds true whether you are dealing with a heath or a lake or a rainforest. In their landmark study of islands, *The Theory of Island Biogeography* (1967), Robert MacArthur and E. O. Wilson even came up with a mathematical formula to quantify the relationship between habitat area and species diversity. Here it is:

$$S = cA^z$$

Where S is the number of species, A the area of habitat, c is the constant and z the slope of the curve when plotted on a log-log graph.

This means that you can calculate the odds for or against survival, as American geographer Jared Diamond did for the birds of New Guinea in the 1970s and 1980s. The main predictor of local extinction, he found, was 'small population size'. There's safety only in numbers. Big numbers. For many island species a population needs to number at least a thousand individuals to be viable. Fall below that and the ecological warning light comes on. The gene pool is closing down; the ecosystem is running empty.

A rather touching example of this is the regent honeyeater of Australia, an attractive black-and-yellow bird with bald patches around the eyes that give it a spurious intellectual look. Once comparatively abundant, the honeyeater has declined through the fragmentation of its habitat to just 300 birds, scattered over roughly 100,000 square miles. This means that it is getting hard for young birds to find a mate. To aggravate their difficulties, some of them are forgetting how to sing. Honeyeaters learn their distinctive song — a series of signature warbles and trills — by listening to others of their own kind. But with so few of them around, they have taken to copying different species of birds instead. That means they have less hope, perhaps none, of attracting mates, and if their mating system is breaking down, then honeyeaters have no future outside captivity (and behind the wire they fail altogether to learn the right song). It seems that a population of even several hundred individuals is not enough to sustain the regent honeyeater as a species.

The relationship between species and habitat area holds the key to successful conservation. Many of the protected sites in lowland Britain are tiny – no more than a few hectares and surrounded by farmland. Unless such isolated sites can be enlarged and linked in some way, protection will probably fail in the end. The small populations of habitat specialists will be subject to the drift of random events, a bad summer, say, a drift of herbicide spray or the withdrawal of funds (or, in rougher parts of the world, perhaps a volcanic eruption or a hurricane). In the end, like Norman Moore's smallest heaths, the special species disappear and the place becomes impoverished and ordinary. This is now widely recognised by conservation bodies whose publications now speak of wildlife corridors and species highways. With land prices being what they are, this is a strategy easier to conceive than to put into practice. One successful example is England's Great Fen project which is gradually stitching together the long-isolated fragments of marshland which are all that remain from agricultural drainage two centuries ago.

The extinction list is full of species that ran out of space. As Charles Darwin found at Galapagos, small islands make wonderful opportunities for evolution, a kind of sealed natural laboratory in which nature can experiment, in some cases wildly. But, by the same token, nowhere on earth is more vulnerable to outside influences. The moment these Edens are invaded by outsiders, led of course, by *Homo sapiens*, the outcome is sad and almost inevitable. By contrast, it is striking how few losses there are among species with large natural ranges. If you wish to survive the Sixth Extinction, therefore, give yourself plenty of room.

Make yourself useful – but not too useful

To survive the Sixth Extinction, it may help to be useful, useful to humankind, that is. The economists who run the world like to put a price tag on everything, and it will obviously help your prospects if you are worth a million or two. It will also help if you are a species celebrity and appear regularly on television. In this world, anonymity only gets you so far. Among the world's lost birds there are species – let's recall the Ascension Island rail, for instance, or the well-named mysterious starling – about which almost nothing is known. They lived and died in the shadows, untouched by scientific enquiry. You might say their extinction was their greatest distinction. But if we have a reason for valuing a species beyond its theoretical right to survive, then a much better case can be made for its preservation, a case that politicians and world leaders can understand.

Animals and plants are useful to us in so many ways. They might be beautiful, or inspiring, or, if you are a plant, provide nutritional or medicinal needs. They might be pollinators (of our crops that is; few care about whether or not they pollinate wild plants too). If a species has economic value, then the balance tilts in its favour. The horse survived as a species because it is fast on its feet and can carry a human rider; if it couldn't or wouldn't, the horse would have been horse-meat, and probably extinct, as it was over most of the world before humankind tamed the last few. Cattle survive because they produce milk in generous quantities and are also good for eating and an excellent use for a field; the camel ditto because it is still the best way of carrying goods through a desert. Britain has a lot of mature oak trees, partly because of their timber but also because they are a long-standing symbol of national steadfastness (for example, the oak is the symbol of the British Conservative Party).

Of course there are limitations. Horses and camels will survive for as long as we cherish them, but that isn't necessarily true of their equivalent wild forms, like Przewalski's horse and Bactrian camel, nor of the European bison, which are among the world's threatened species. The ancestor of the cow, the mighty aurochs, has been extinct since the seventeenth century – and most of the other wild cattle of the world, kouprey, gaur, banteng and the wild water buffalo, are heading the same way. Both the European and North American bison were saved from extinction at the last minute, but only by semi-domestication. For fully domesticated animals, like the llama or the dromedary, there are no wild forms left. They belong to us now and humankind is their destiny.

In theory all species are equal and all have a right to exist. In practice, we spend the bulk of resources on animals we like. We can easily manufacture reasons for doing so. In Britain, nesting ospreys and sea eagles attract tourists and hence income to places that need it. The beaver has been sold as an animal with a job to do, improving water quality and preventing floods. You could say the same about the wolf, which, after its reintroduction to Yellowstone National Park, has done the Park a service by weeding out weak and sickly deer and lowering the deer population overall. This, in turn, has led to greater woodland regeneration and less erosion on the river banks. In North America supreme efforts were made to prevent the extinction of the whooping crane, the trumpeter swan and the Californian condor, all large and prestigious birds. Effort delivers value and it seems that species which are drawn back from the brink by human intervention matter to us more than those which recovered on their own. In Britain, for instance, there were plenty of cheers for the carefully planned and monitored progress of the introduced beaver, but very few for the wild boar, which simply jumped the fence. There's a feeling that the boar isn't quite playing the game; that its self-introduction denies us the sense of control that, consciously or not, seems to be a basic human need.

Perhaps the clearest case for usefulness can be found in the contrasting fates of two related birds on opposite sides of the Atlantic, the British red grouse and the American hazel hen (not to be confused with the European hazel grouse which is a different species). The red grouse is a useful bird. As a target, and as food, it is prized above all others: avian caviar. Grouse shooting provides income for moorland owners, and repays the effort invested in gamekeeping and heather burning to maximise the number of birds the land can hold. It has a close season before the massive shoot on the 'glorious' 12th of August when, coincidently, the heather is at its brightest. Shooting, ironically, is what keeps the red grouse off the Red List. Without management it would probably be quite scarce and retiring, like its continental confrère, the willow grouse.

The extinct hazel hen also tasted good, but in its case there was no regulation, no bespoke habitat management and no close season. They were simply shot until there were no more to shoot. In Britain grouse shooting is an event, part of the social calendar of the wealthy. Not so the hazel hen which probably owed its sad fate to the right of the American people to keep and bear arms, a sacrifice to the Second Amendment of the US Constitution and to competition from our far more useful flocks and herds.

On the other hand, don't be too useful . . . as whales have shown, their usefulness nearly proved their undoing. On a large ship with the right equipment, a 50-tonne whale could be stripped down to 50 tonnes of useful product: meat, oil from heating the blubber and baleen for making 'whalebone' corsets. The sperm whale additionally gave us ambergris, used in the perfume industry. Whale oil burns slowly and without an offensive odour. It lubricated the cogs of the Industrial Revolution, and gave us superior brands of soap, varnish, cosmetics and paint. The problem was that there weren't enough whales. In the 1920s, when industry was reaching new heights, with bigger ships and more powerful harpoons, they targeted the biggest of them, the blue whale. A single blue whale could be melted down into 120 barrels of oil. When blue whales became too scarce to hunt anymore, they turned to fin whales and when they, too, were hunted out, to smaller sei whales. By the time whaling was banned in the late 1980s (whale oil no longer being much used), they were down to the much smaller minke whale. Usefulness in their case caused rather than prevented near extermination. Commercial whaling has more or less ceased (though at least three nations would very much like to renew it) but, of course, whales face new threats now: plastics, collision with ships and the continuous roar of engines. In the modern world, no big species is safe.

Nor is usefulness helping elephants, rhinos or pangolins. *Au contraire*, they are the most poached animals on earth: living repositories of ivory, horn and scales. The trafficking of the world's pangolins, which has already driven one species close to extinction in China, is almost too much to bear. The value of this small, defenceless animal lies mainly in its coat of scales, used in traditional medicine. The trade is technically illegal but pangolins, live or dead, are nevertheless on open sale in Eastern markets. The terrified animals are bundled into small crates, transported across the world, then slaughtered in a crowded market, parted with their scales (one hopes after they are dead), taken home and eaten. If the pangolin really did spread Coronavirus, as has been suggested, in its blood, urine, shit, in its last terrified breath, then we have all paid the price: the pangolin's ultimate revenge, comeuppance for *Homo sapiens*.

Keep your head down

As Carl Sagan once informed us, in the long-term, 'extinction is the rule. Survival is the exception.' (Sagan 2006). The price you may pay for being an exception is an exceptionally dull life, the sort of existence where nothing sees you and you see nothing except the mud in which you are probably living. There is a living animal called Amphioxus or the lancelet, a kind of proto-fish, which closely resembles another creature preserved in rocks 500 million years old called *Pikaia*. To transform from *Pikaia* into Amphioxus is not a very big leap. Evidently their lifestyle didn't require much adjustment over all that time – the entire history of advanced life on earth. Amphioxus survives by literally keeping its head down, half buried in the mud, filtering grot from the water. But if extinction is the worst thing that can happen to a species, then Amphioxus does show us one way out.

Species of mammal, on the other hand, last only one or two million years on average. Our relatives, the Neanderthals, managed only a few hundred thousand years, a wink in the eye of eternity. Of course the Neanderthal may well have been killed off prematurely after it met *Homo sapiens*, a story re-enacted in William Golding's novel, *The Inheritors*. But the general point is that sophistication, existential fine-tuning, comes at a price. Environmental change hits big species harder than small ones. A modern Aesop would make a fable from it: immortal mice laughing at ephemeral leopards.

Is there another way of keeping your head down, by avoiding humankind altogether? Maybe, but those refuges are disappearing fast. The last unexplored vista of the world was the deep sea. The fish that lurk in the dark ocean depths are mostly what ecologists call K-strategists, meaning that they respond to an unchanging but nutrient-poor habitat by slow growth, late maturity, a long life and rela-tively few babies. The orange roughy, for example, does not breed until it is twenty years old but lives well into its seventies or even longer (much longer for the lucky few; the age of one elderly roughy was estimated as 230 years). That strategy served them well so long as they kept out of the way of trawler nets. But unfortunately trawling is no longer restricted to banks and coastal shelves. Nets can now reach 2,000 metres depth, more than a mile down, and they not only hoover up the fish but also destroy their nurseries. In 2008, 27 countries were using deep-sea trawl nets across all the world's oceans. A study of five species commonly caught up in those nets by the Canadian Department of Fisheries found that all five had crashed. One species, the roundnose grenadier, had declined by 99.6 per cent within a quarter of a century. The other four were thought likely to follow it into oblivion by the end of the present century, and even that timeframe might be optimistic. Numbers of deep-sea gulper-sharks, noted for their oily, protein-rich livers, crashed after only two or three years of targeted fishing, and they too are now in the critically endangered category. The trouble is that, unlike cod or herring, these slow-breeding deep-sea fish cannot replenish their numbers quickly. They live in the slow lane of long lives and low fecundity. Deep-sea fishing is manifestly unsustainable, closer to mining than conventional fishing, and one day quite soon there may be no more fish down there worth catching. Of course the world is not short of good advice on sustainable fishing, and most countries have their own fishery regulations. Does it make much difference? As far as I can ascertain, the answer to that is long and complex, riddled with ifs, ands or buts, yet concludes with one word: no.

Try to stay cheerful

Nearly a hundred years ago, an American scientist called Warder Clyde Allee came up with the novel idea that small populations cause a kind of species depression. In so many words (actually quite a lot of words), he believed that low numbers result in low spirits. A homely example would be a goldfish in a bowl. It seems that a lone goldfish meandering pointlessly round and round will die sooner than a group of goldfish swimming about in a nice big tank. You might reason that loneliness leads to self-neglect and a premature death. (Having once kept a goldfish I had accidentally won at a fair, I find that idea very plausible, especially after, getting bored with its antics, I tipped it into my grandmother's bird bath.) According to Allee, this is true of whole populations too. Species in general find life easier and more pleasant in large numbers. When populations fragment, the pressures pile up and life gets tougher and tougher. Perhaps there comes a tipping point when a species effectively gives up? Sometimes it seems that way. Look at the above-mentioned hazel hen, once present in North America in the millions, then mere thousands, then hundreds, then scores, until all the care in the world could not prevent a total collapse resulting in just one last lonely bird, called Booming Ben. He died in 1933. Perhaps in some undefinable way, hazel hens failed to cope with stress, decided that life sucked and eventually stopped trying. On the whole, I find this explanation less than plausible.

Some birds can and do recover from very low numbers. The classic example of such a comeback is the Chatham Island robin. Confined to one small island and nearly exterminated by cats, at its lowest point the robin population was reduced to just five birds, including only one fertile female called Old Blue. They were rounded up for last-minute captive breeding using similar-sized tits as foster parents. And now, forty years on, there are 250 robins – though, alas, no longer living on Chatham Island – all descended from Old Blue. Of course, they are inbred, but for small island birds like this which may well have experienced such genetic bottlenecks before, low genetic diversity may be something they can live with.

At the other extreme is the famous North American passenger pigeon, perhaps once the world's most abundant bird (if tales of them darkening the skies for miles are true). Yet in little short of fifty years, the species plummeted from possibly billions to just one, a captive bird called Martha, who died in 1914. It was the biggest avian species crash in history. Mark Avery, in his book, *A Message from Martha*, fingers habitat destruction as the primary cause. The pigeon depended on large woods with their copious but only occasional crops of seeds and nuts. To find enough to feed those enormous flocks, they had to keep moving on. But by the mid-nineteenth century half of those woods had gone and formerly large areas of natural forest had fragmented. The pigeon spent more calories finding food than before, efforts that probably slowed its reproductive rate, while all the while it was being trapped and shot in great numbers (after all, you could hardly miss when there were birds enough to darken the sky).

Presumably there came a tipping point when the pigeon's life strategy, its reliance on huge numbers, no longer really worked. Anyone watching those winter murmurations of starlings cannot but be impressed by the uncanny ability of the individual birds to think and fly as one, turning into a kind of super-organism. Perhaps the passenger pigeon, too, functioned only as a unit in a vast cloud of birds. Did this uncountable mass, this corporate super pigeon, have senses beyond that of individual birds? And if so, did it have some premonition of what was happening and what lay in store for it? For Mark Avery, the message from Martha, the last passenger pigeon on earth, is that it could happen to any bird, anywhere. Life's a lottery. Perhaps it could happen to *you.*

Personally I think she had a different message. 'Whoever you are, stay upbeat and cheerful. Don't be a loser like me, be an optimist. Above all don't let humankind get you down'. And with that, Martha toppled off her perch.

> *Hop and skip to Fancy's fiddle,*
> *Hands across and down the middle –*
> *Life's perhaps the only riddle*
> *That we shrink from giving up!*

Chapter 5

Paths of extinction

The eagle owl is regarded as a bird of death ... It inhabits desert places that are not only desolate but also terrifying and remote. It is a monster of the night, and its voice is not so much a song as a deep groan. So if it ever appears in a city by day it is a dire portent.

Pliny, *Natural History* X, 34-35.

Extinction always has a date, though the exact moment marking the death of a species will nearly always be unknown. As we have seen, it may take decades to be sure whether an animal or plant or insect really has vanished forever. And even then, certainty would depend on an adequate database of records. Hence in the overwhelming number of cases extinction has to be deduced retrospectively. For invertebrates, the IUCN insists on a time lag of at least fifty years between the last record of a species and a declaration of extinction. And it tends to qualify even that with the word 'probably'.

A good example of this extinction time lag is the Caribbean monk seal, the only large North American animal to become globally extinct within the last sixty years. Monk seals are so named because of the characteristic fold of fat on their necks which resembles a monk's cowl. This particular creature was large, up to eight feet long, with a round, wide-eyed, trusting face. It was declared 'probably extinct' in 2008, after an exhaustive search, lasting five years, but had been on the endangered list for decades, leaving plenty of time, you might think, for some sort of recovery plan to be formulated. But nothing happened, and for good reason. The IUCN were dealing with a phantom animal. The seal had already gone. The last confirmed sighting of a real live Caribbean monk seal was in 1952, fifteen years before it was officially declared endangered, let alone extinct. In truth, it was probably doomed long before that. Seals live a long time, and there would have come a point when the harassed, starving animals were no longer breeding successfully. In statistical terms, when recruitment of young lags behind mortality, the survival curve dips down and the species is in deep trouble. A time will come in the biological lifetime of any species when, faced with numberless stresses and strains, extinction becomes the most probable outcome. In the case of the Caribbean monk seal, it just came sooner than anyone expected.

The loss of the monk seal entailed the extinction of another species that depended on it. Inside the warm, moist nasal passages of the seal lived a mite, *Halarachne americana*. Apparently, no other kind of seal's muzzle would do: it was the Caribbean monk seal or nothing. One can imagine the things crawling out of cold, dead seals and dying of despair. Perhaps every mammal on earth has some unprepossessing parasite that depends on it for survival (we certainly have some: *Homo sapiens* is among the most parasitised animals on earth). Presumably there were other dependencies too.

For example, the sharks that used to prey on the seals might have gone hungry.

Each animal or plant extinction has its own unique pathway, its own trajectory of descent. It will, of course, only be fully known for those species for which humankind has taken a special interest. For the Caribbean monk seal, it seems to have been a fatal combination of tameness, utility and living in the wrong place. The live seal was packed with useful blubber-oil that could fuel lamps and waterproof the hulls of boats, while its meat made a stew rich in Vitamin C and omega fatty acids. You could preserve its skins to make trunk-linings, coats and bags. Like a pig, you could use practically everything bar the squeak. Slow and lethargic on land, the seals would watch with their large, curious eyes as their killers approached. They would witness their fellows getting shot and whacked, and perhaps wonder what was happening for their evolutionary experience had never prepared them for such horrors. Their identified enemy was sharks, not mysterious figures on two legs. Rocky islets, surely, were safe? Later on, over-fishing of their Caribbean reefs – for unfortunately the seals liked to eat the same fishes and molluscs as we do – led to starvation and reduced the number of pups the underfed seals could support. Oblivion overcame them just as people were starting to ponder those novel ideas, sustainability and conservation.

And so we know in broad terms what happened to the Caribbean monk seal. It even has a kind of afterlife, a past worth studying for the light it sheds on conserving stocks and on how we might still manage to save its critically endangered fellow monk seals in Hawaii and the Mediterranean. It has a retrospective biology too. DNA extracted from preserved skins in museums suggests that our seal was more closely related to the Hawaiian than the Mediterranean one. Some say it should now be placed in a new genus, *Neomonachus*. Rather ironically, this means 'the new monk'.

Since extinction is a scientific topic, it is generally discussed in scientific terms: formulae, models, graphs. But where extinction becomes a matter for public concern and debate, as it has, the science becomes infused with cultural concerns: of human responses to loss, of guilt, and the hope that something worthwhile can be learned from the experience. In general, and disregarding bugs and viruses that harm us, we do not want species to die out – but if they do we want it to mean something. The cold voice of science gives way to emotion and the facts, you might say, begin to warm up.

It is in that thought that I want to tell the stories of some lost, or nearly lost, species and the meaning that one can take from their disappearance. One, the seemingly trivial extinction in Britain of a small grey moth, says something, to my ear at least, about the unexpectedness, the apparent randomness, of loss. Extinction has a habit of constantly catching us unawares. The second is a fishy tale, which, I think, invites the question of whether it is right and proper, or even honest, to try and 'reintroduce' a species once its natural environment has been destroyed. The third offers the loss of the wolf, which left Britain and Ireland without any big, dangerous animals, to ask whether there are limits to our evident yearning for wild places. The fourth is about the Yangtze river dolphin or baiji, the first species of the world megafauna to die out in the twenty-first century. Its fate invites us to ponder what life was like for an intelligent species in its final days. It is not a pretty story. In fact it is heart-rending. But it offers food for thought.

Tales of lost fishes

'Ugliest fish I've ever caught,' recalled an unnamed angler. 'Slimiest too' (Westwood & Moss 2015).

It was a burbot, and it is, indeed, a weird looking fish: long and thin, apparently without scales, like a plump eel, but endowed with a flattish, snake-like head and a mouth as wide as a toad's. Long fins extend down both sides of its slippery body culminating in a lozenge-shaped tail. The burbot's mottled colours have been described as blotched or even 'fungoid', as though the poor fish was suffering from some terrible affliction of the skin. When caught it tends to twist and squirm and try to anchor its long body around one's arm or leg. The name burbot comes from *barba*, the Latin word for a beard (and so in modern French, which is closer to Latin than English, it is a *barbot*). The fish's 'beard' is its barbel, a sensitive, fleshy projection attached to its lower lip like a goatee. It helps the fish find its way along the river bottom in the dark.

I first came across the burbot not in a river but on a cigarette card. It was card no. 8 in a series of 'Fresh-water Fishes' issued by Players cigarettes back in 1935. This burbot is exiting serenely from a sketchily drawn patch of weed, a finny, mottled-brown torpedo. It is 'an exceptionally greedy fish', notes the card; it 'eats just about anything'. And we in turn will eat the burbot because 'its flesh is excellent'.

The next time a burbot swam into my life was on another picture card, this time issued by Brooke Bond tea back in

1960. This was a livelier creature, proceeding upriver with a flex of its long, curved rear and with a don't-mess-with-me look on its face. We learned from the text on the back of the card that 'it is found, not very often, in a few Eastern rivers', lurking under stones or among tangled roots by the bank. There were, it seemed, 'no angling records'.

As a little boy, I liked the look of the burbot and would eagerly ask my grown-up angling friends where we could find one, and if I could take it home. In fact none of them had ever heard of it. And no one at all has caught a burbot in Britain since 1969. The very last one, about a foot long and weighing a modest pound, was caught that year by one John Dean, one September night, on the Old West River, part of the Great Ouse system, near the fenland village of Aldreth. It now resides, head upwards, in a pickle jar in the Cambridge University Museum of Zoology: the Last British Burbot.

Time was the burbot was quite a famous fish. You can tell that from its Latin name, *Lota lota*, which literally means 'fishy-fishy', from the Old French word for a fish, *lotte*. *Lota lota* is the defining fish of the cod family, the Lotidae, of which the burbot is the only freshwater member. To put it another way, the burbot was once better known than the cod! There was even a pub in the Cambridgeshire fens called Pout Hall named after one of the fish's folk names, the eel-pout. Other folk names, all signs of past familiarity, included coney-fish (that is, rabbit-fish, presumably from its habit of hiding in holes in the bank), lawyer-fish (it shared the lawyer's traditional goatee beard), lingcod or freshwater-ling (from the resemblance to the salt-water ling fish), and mudblower or mother-eel. As you might expect from a relative of the cod, it tastes good, high in protein, fleshier and less bony than most freshwater fish, and with a sweetness that spawned yet another nickname, 'poor man's lobster'.

The burbot has a literary history too. Aelfric, writing a thousand years ago, mentions a strange fish that sounds like a burbot. Leonard Mascall's sixteenth-century *A Booke of Fishing with Hooke and Line* notes that it was then common enough to be fed to pigs. The burbot was the piscine hero of Chekov's short story 'The Fish', while burbot soup makes a dish fit for the Tsar in Tolstoy's *Anna Karenina*. In the Great Lakes, in Alaska and in Norway, they still fish for it under the ice around spawning time in late winter when the burbot is uncharacteristically lively.

A century ago, burbot still lurked in many eastern British rivers, although it was most at home in the slow, deep, weedy rivers of the fens. It is a cold-water fish, laying its stream of tiny, oily eggs – half a million of them per kilogram of body weight – in the dead of winter in water temperatures as low as 0.5 degrees Centigrade. Such cold fish depend on clean, well-oxygenated water. The present-day lack of cold, clean, deep, river water is almost certainly the reason why there are no more burbots in England. It was last seen in the River Nene in 1904, the Blythe in 1933, the Waveney in 1964, the Trent and the Yorkshire Derwent in 1965, the Cam in 1969 and, as mentioned previously, the Great Ouse in that same year when Mr Dean caught the last one.

Even if the burbot were still around in Britain, however, where would it reside? Just look at the state of rivers now. In 2020, the Environment Agency rated the Cam as 'poor to moderate', the Nene as plain 'poor' and the Great Ouse as 'bad', that is, even worse than 'poor'. Only 14 per cent

of British rivers met the Agency's criteria for 'good ecological status', and none of those were former burbot rivers.
Every main waterway in its old stronghold in Cambridgeshire
is chemically polluted; there were seventeen recent agricultural 'incidents', such as raw slurry pouring into the
water, in the lower Cam basin alone.

'Poor ecological standards' means, among other things,
a lot of mud, little weed, warmer water and low levels of
dissolved oxygen. All of which militate against the survival
of fish like the burbot. The *Angling Times* offered £100 to
anyone who could find a burbot anywhere in the UK;
money that has not yet been claimed (or rather, there were
claims but no firm evidence by way of photographs or
actual fish). So the burbot is officially extinct in Britain:
in fact, it is our only recently extinct species of freshwater
fish.

Further back, in pre-industrial times when estuaries held
clear, healthy water and you could see through the ripples
to the gravel beds far below, there were other fish, now
lost. The noblest of them was the sturgeon. British sturgeon
appeared on menus in the Middle Ages – it was a royal
fish, reserved for the king's own table – and it used to
spawn in the lower Thames, close by Westminster and
Whitehall. Sturgeon are still caught in the North Sea by
trawlers from time to time, but they seem to be wanderers.
It hasn't spawned in British waters for a long time and is
deemed to be nationally extinct as a breeding species. As
is a much less exalted species, the houting, a kind of small
herring with a curiously projecting lower lip. Like the
sturgeon, it spawned in clean estuaries facing the North
Sea, but not anymore – and it is, indeed, in serious decline
in the equally polluted estuaries of the Rhine, the Weser

and the Elbe. Off you go, little houting, a fish no one knew, and so no one cared about. Its only contribution to our kitchens was on another card issued by Brooke Bond tea.

The houting has some still living but landlocked relatives usually grouped together as 'whitefish'. They have been given different names from place to place, *gwyniad* in Wales, *schelly* in Cumbria, *pollan* in Ireland, *powan* in Scotland, but there are probably only two true species, the whitefish proper and the vendace. All of them are threatened by pollution, climate-induced changes in the water and from competition by introduced fish. Fish used to living in isolation become vulnerable when humankind tips the scales. Loch Lomond, for instance, is home to the powan which has probably been there since the last Ice Age. Unfortunately, anglers decided to release there a more useful bait fish, the ruffe, a smaller relative of the perch. And all too predictably it has proceeded to out-compete the powan, not least by devouring its eggs, and replacing it as the principal fish of the loch.

For isolated freshwater fish, introducing foreign species is a threat to their survival. Scale Loch Lomond's powan problem up a thousand times and you are still not even close to what happened at Lake Victoria, the world's largest tropical lake. Lake Victoria and its fellow deep-water lakes in Africa's rift valley have, or had, the richest biodiversity of freshwater fish in the world; each lake packs in more species than there are in the whole of Europe. Four out of five of these fish are cichlids, a group which evolved explosively in those inland waters producing hundreds of species. Many of them are colourful and are the freshwater equivalent of the fish of coral reefs (and they are equally tame and follow the diver). Lake Victoria alone had 500 species of cichlid and counting (for new species are discovered almost yearly). And then, in the 1950s, they introduced the Nile perch.

Nile perch are nutritious and good to eat; they also grow very large and are popular with sport anglers. Unfortunately they are also one of the world's worst invasive species. To cut a long story short, Lake Victoria is now short of at least 200 species of native cichlid and many of the others are in danger. Before 1950, nine out of ten fish there were cichlids. Now it is more like one in a hundred, and the perch has replaced the cichlids as the lake fishery. It wasn't a simple matter of the big fish eating the smaller ones so much as competition for resources. The Nile perch has teeth and jaws to match its appetite. But the cichlids, by a quirk of evolution, have less efficient toothless jaws that act more like meat tenderisers. They take much longer over a meal than the perch. And there was another unforeseen ecological consequence, this time on land. The Nile perch is a fatty fish and must be smoked to avoid spoiling in the heat. And that led to an increased demand for wood in an already deforested region. So forest species suffered too. We know not what we do when we mess with nature; but we should, because what habitually results is chaos.

Compared with the ecological catastrophe that has overtaken the cichlids, of which over a hundred species are probably globally extinct, the loss of just one fish, the burbot, might seem trivial. All the same, it is a high-profile fish and a most distinctive one, and some people would love to bring it back to Britain. A successful reintroduction would depend on a number of 'ifs'. If suitable stock can be located. If a clean and suitable river can be found. If someone agrees to fund it. But, you have to ask, what would be the point?

For George Monbiot, who championed the burbot for the Radio 4 series, *Natural Histories* ('25 extraordinary species that have changed our world!'), there is sufficient reason in simply knowing it was still there, somewhere down among the weeds, 'something you can't put a price on'. Jonah Tosney, a fish expert, agrees, pointing out, 'It's a native species. It *should* be here.' But Alwynne Wheeler, former curator of fish at London's Natural History Museum, is not so sure: 'If it's simply because they used to be there, that's not a very good argument. It would cost a lot of money and also wouldn't be very popular with anglers as the burbot would eat the eggs and young of salmon and trout'. (Yes, but the salmon and trout would also eat the eggs and young of the burbot! It's a jungle out there.)

As it happens, there is, indeed, a costed plan to reintroduce the burbot to Britain, masterminded by the Norfolk Rivers Trust with the help of a grant from Natural England. First there will be DNA sampling to ensure that there are no unseen burbots already lurking in the rivers. Then captive-bred stock from European hatcheries will be raised and released, though the release sites had not been divulged at the time of writing. The Trust points out that the burbot has been reintroduced into rivers in Belgium and Germany, with apparent success. So, why not here as well?

Why not? Well, let's put it this way: the burbot's tale, though undoubtedly tragic in conservation terms, has meaning. Its extinction is a sounding bell, reminding us of the sorry state of our watery environment. It tells us, in effect, that Britain is no longer a fit place for burbots. It cuts through the nonsense, offering proof that something has gone drastically wrong and that lowland rivers are so changed from their natural state that they can no longer support their former diversity of fish. To be blunt, the waterways of the fen country are no longer living rivers so much as muddy troughs. Even supposing that the burbot could live again, lurking contentedly in some duly sanitised pond or stream, where is the public good? To pretend that its extinction was only a temporary blip, and that matters can be put right by reintroducing the species from somewhere else, is surely to deny extinction its power, its imminence. It feeds into our false sense of certainty, that with enough goodwill and resources, we can always fix things: fish back, problem solved. Bringing back the burbot might make us feel masters of events. But are we? I think that returning it to good old British mud would misread the message. That ugly fish was trying to tell us something.

A lesson from the moths

One night during the Proms of 2013, a co-production of the BBC and the Danish National Vocal Ensemble performed Harrison Birtwistle's latest piece, *The Moth Requiem*. I watched the concert on YouTube. Birtwistle's music is, you might say, an acquired taste. But while, to my untutored ear, most of his output sounds like pops, whistles and bangs, the twenty-minute *Moth Requiem* is Birtwistle at his best: not exactly tuneful, never that, but cumulatively mesmeric, quite inexpressively mournful in its softly wailing choir accompanied by a warbling alto flute (during which three harps imitate the sounds of an unlucky moth caught inside a piano). Birtwistle, you think, must have been really moved by the passing of a moth.

The composer had real moths in mind. He had raised silkworms as a boy and liked moths in general for their beguiling mystery: their shadowy lives out in the dark, broken by the brief moments when they accidentally enter our own living space. They flutter and glow at the window, eyes shining gold or red, and then are gone again, slipping back into the void. Birtwistle was entranced by the poetry of their names. Whoever called them clouded brindle, marbled carpet, cistus forester, nut-tree tussock or green-brindled crescent must have had a sense of poetry. Some British moths are survived only by their names for they seem to have vanished. If you listen very carefully, with the sound turned up, you can hear the Birtwistle choir intoning some of these names: frosted yellow, dusky clear-wing, scarce dagger, union rustic, Lewes wave (yes, we've waved goodbye to the Lewes wave), all gone, perhaps forever. The *Moth Requiem*, explained the composer, is a meditation on loss, a memorialisation of beings that will come no more, treated with the seriousness that Mozart gave to a requiem mass. The lost insects were, to the composer, 'an emblem of things disappearing', of the death of a species and the finality of extinction.

According to Butterfly Conservation, sixty-two species of moths died out in Britain during the twentieth century. Only moth experts will have noticed, and even they do not know why these particular moths are no longer with us. We don't know why the union rustic, for instance, died out, only that it did. All that is left of it is a tiny insect footnote in ecological history, a faded line in the fat ledger of lost species.

I once held one of these lost moths in my hand. It was a bordered gothic, *Sideridis reticulata* ('iron-form, netted').

It was slurping the mothing treacle I had thoughtfully brushed onto the trunk of a pine. Like so many moths, the bordered gothic is subtly pretty in its mottled dead-leaf camouflage. Its name recalls the pattern of fine white lines on its dark wings, as if drawn by a pen and resembling the Gothic tracery of church roofs. It is distinguished from another moth called simply The Gothic by a dainty border of dashes and spots, hence *Bordered* Gothic. And there it was lapping up the treacle, oblivious to the red shine of my torch (for moths can't see red). More bordered gothics appeared from the depths of the pines and settled on the sheet spread out beneath our glowing lamp. As each new species flew in, my friend solemnly intoned their names: bordered gothic, clouded border, dark brocade, light arches, burnished brass . . . oh, and another bordered gothic.

There were, we thought, more exciting species about on that warm night in the Norfolk Breck. Back then the bordered gothic was considered to be a widespread, if not very common, moth. In the first Red Data Book for insects, published in 1987, it was not listed. Yet, by the time the *Atlas of Britain's and Ireland's Larger Moths* was published, in 2019, our little moth had declined so far as to be considered 'extinct as a resident in Britain'. 'Resident' means a permanent presence, a species breeding year-on-year on British soil through a full cycle of eggs, caterpillars, pupae and emergent moths. The few bordered gothics reported in recent years seem to be continental strays along or near the East Anglian coast. It still survives, precariously and as a distinct subspecies, on the southern Irish coast.

The bordered gothic was not among the sixty-two species of moth lost during the last century. It is an up-to-date, twenty-first century extinction, and it has already been joined by more departing moths: orange upperwing, Essex emerald, Brighton wainscot, stout dart (the latter being another moth I'd encountered back in the 1980s, sleeping through the hottest days of summer in a friend's garden shed – the moth, I mean). What ailed these lost moths? Why them? On the face of it, the bordered gothic did not seem to be a particularly fussy species. Its food-plants include campion and knotgrass, both common wayside flowers. Its past haunts included field margins, chalk pits, disused quarries, embankments, cliffs, all features of a mellow, well-used landscape – but perhaps hinting that it would be grateful for warm, dry, open environments, not so much for itself as for its more vulnerable caterpillar.

Unless you are a regularly monitored species like birds or butterflies, extinction creeps up on you without warning. For the bordered gothic our perception of its decline to national extinction was not so much an observed process as a changing of labels. Noting a dearth of recent records, the moth was made a 'Priority Species' and added to the Biodiversity Action Plan list. But no funds were available and so there was no action. The Butterfly Conservation charity (which also deals with moths) decided to place the moth on its list of the 'Nationally Scarce' species, and also considered adding it to the Red List, the top tier for troubled life. Once on the Red List, any species clean out of luck is likely to be promoted from 'rare' to 'vulnerable', and, if the decline continues to the danger point, to 'endangered'. The step after that is 'extinct'. And so it proved.

Since biodiversity action began in 1992, scores of British insects and spiders have received action plans. But in the case of a little-known species like the bordered gothic, what possible action could there be? The reason for its decline, noted Butterfly Conservation, 'is not fully understood'. That is something of an understatement, for it is not understood at all; the moth was never studied. All we know is that it was a species you would come across now and again when running a trap (or painting the trees with treacle). And now you didn't. Extinction snapped at the bordered gothic out of the blue, and out of sight. Perhaps the next time we hear about this quietly pretty little moth will be when Harrison Birtwistle writes *Moth Requiem II*.

The extinction of just one insect is hardly a cause for panic. We miss the song of the nightingale and pity the poor, rare animals we see on television, but I probably mourn alone for the bordered gothic moth. But, more seriously, moths in general – and in Britain that is around 2,500 species, and worldwide a truly vast number – are failing to respond well to the modern landscape. The most objective measure for how they are doing is the Rothamsted Insect Survey, a nationwide network of light traps which have been run by volunteers since 1968. There are currently about 80 such traps across Britain and Ireland, and taken together, their data suggest that moths have decreased by an average of 28 per cent across Britain, but by as much as 40 per cent in the south. Even this may underestimate the real decline. Drivers of a certain age will recall the 'moth snowstorm' that could appear in the headlights on warm, still nights. The nights are just as warm now, or warmer, but those mothy blizzards are a thing of the past. How often now are we bothered by a buzzing moth? Moth hunters no longer search posts and tree trunks as they used to because the moths just aren't there anymore. Based on a survey of 337 widespread moths, Butterfly Conservation discovered that two-thirds of them have declined since 1968, some quite severely – and at least one, the V-moth, has passed, with some speed, from a garden pest to 'endangered'. In this case, 'V' obviously does not stand for Victory.

The decline of familiar species is more disconcerting than rare ones because it is not expected. Think of the disappearance of sparrows in London or the sudden, out-of-the-blue dip in numbers of tortoiseshell butterflies in recent times. When common things disappear you can be sure that something broad-ranging is happening, something bad, some thinning of the environmental fabric of which the sparrow or the butterfly are only the outward signs.

Similarly, the decline of moths across the entire range of species, common or rare, suggests an underlying sickness. Sir David Attenborough viewed the moth-trap evidence as 'significant and worrying', and 115 MPs agreed with him sufficiently to sign an Early Day Motion in the House of Commons. Moth decline contributed to the mounting evidence that eventually resulted in the banning of neonic pesticides, the surest insect killers ever devised.

There are some reasons to be more cheerful. In island Britain, as I've noted, sixty-two former resident moths have died out, but at least twenty-seven new ones have settled, and over a hundred more have appeared for the first time, perhaps in response to a warming climate. Potential new colonists are arriving at the rate of two species per year. The old moths had grown up with the British rural landscape over centuries, finding their niche and sticking with it as the tide of progress ebbed and flowed around them. But some of the new ones are townies: tree lichen beauty in London gardens, small ranunculus on allotments by the Thames. We shall have to wait and see how they do. The successful ones may be those that can best take advantage of urban, human-made opportunities: weedy ground by shopping malls, perhaps, or roof gardens and wall tops, urban greenings. Good luck to all of them.

The moths mourned by Birtwistle's dirge belong to bygone generations, the rural landscape of Edward Thomas or Thomas Hardy. Evidently every one of them possessed some flaw in their make-up, some lack of adaptability that proved fatal. We are poor at predicting extinctions. Even moth experts tend to assume that the rarer the species, the more vulnerable must it be. I think I disagree. If I was looking for the next extinct moth, I wouldn't be looking at the rarest of them. I'd be perusing the list of species which over-winter as caterpillars. As winters get warmer but wetter, their strategy of sleeping through the cold weather won't work. The caterpillars will wake up, wander about, lose food reserves and energy and starve – and that's if mildew or disease or a hungry bird doesn't get them first. If our lost moths tell us anything, it is that extinction can be almost terrifyingly random. Send not to know for whom the bell tolls. It could be any species, at any time. There is no justice in extinction. It is as remorseless and indiscriminate as the plague. It continually takes us by surprise. Look and it is there. Look again and it has gone.

Call of the wild: rewilding and the British wolf

There was a time, a distant time, before the ice sheets obliterated everything and wiped the slate clean, when ancient Britons lived with dangerous animals. There were a couple of elephants, to start with, plus bears, hyenas, big-horned, bad-tempered bison and wild cattle, and even (if you go back far enough) cave lions, bigger and heavier than their African equivalents. Our ancestors lived within a fully functional ecosystem. They were, and no doubt felt they were, part of nature – and a very uncomfortable existence it must have been when you consider that they were surrounded by species that could, and no doubt did, raid their food-stores, or trample down their huts, or even have them for lunch. Humankind was of course illiterate and scientifically ignorant, as well as scared. The only historical evidence from that distant age are some scratches on cave walls and bare bones: enough bones to know that man and hyena, for instance, lived in close proximity. People must have been constantly looking over their shoulder. Out on a walk they would have kept a firm grip on their bows and spears.

The hyenas didn't survive the last Ice Age. Wild bears still lurked in forests in Roman times, but they never became part of any literary tradition (later dancing bears and baiting bears were either captive bred or imported). Aurochs morphed into cattle or disappeared. Wild horses were replaced by tame horses. The only big fierce animal left – or at least the only one big enough to bother us – was the wolf.

The British and Irish wolf is, of course, extinct, and some of us would love to have it back. We know from its bones that our particular race of wolf was a fine animal similar to arctic wolves with their thick, winter-grey coats. It once roamed all over Britain and across Ireland too. It was part of our history. It seems we admired, feared and hated it – and eventually, of course, we exterminated it.

A fierce and cunning pack animal, the wolf was the sworn enemy of farmers. Its presence meant that cattle and sheep had to be guarded, especially at night, which cut into a landowner's income. But wolves could be hunted and trapped, and they were gradually pushed to the margins of cultivation. The agricultural lowlands were probably already wolf-free by Saxon times, to judge from the rarity of wolf place-names in Kent and East Anglia. Later on, the forests and borderlands and uplands were gradually cleansed of their wolf packs too, as were the once thriving wolves of Ireland. You might say that wildness and wolves went together and vanished together. By around 1500 in England and Wales, by 1700 in Scotland, and by 1800 in Ireland, the whole land was free of wolves. And no one seemed to mind a bit.

It seems as though the Anglo-Saxons, at least, perceived nobility and grace in the wolf. Look at some of the names bestowed on aristocratic children: Beornwulf ('wolf-man'), Wulfnoth ('bold-wolf'), Wulfgar ('wolf-spear'), Wulfheard ('hard-wolf') – warrior names, borrowed from the fiercest beast available. There were at least two wolf-kings: Ceolwulf ('ship-wolf') and Aethelwulf ('noble-wolf'), who was the father of King Alfred ('wise-elf'). There was even a wolf-tribe, the Wuffings, descendants of one Wuffa. Is it a consequence of our lost wolf-consciousness that few, if any, children today are given wolf names?

There are also at least two hundred place-names that recall the former presence of wolves, scattered across the land from Cornwall to Sutherland: Woolner, Woolpit, Woodale, Wolfheles (but not Wolverhampton, which is Wulfrun's High Town). It seems that when wolves were still around, people were very conscious of them. In the tenth century, King Athelstan (though some say it was King Edgar) demanded an annual tribute from the ruler of Wales of 300 wolf pelts annually. Evidently, then, there were plenty of wolves in Wales. After a battle, wolves were said to join 'white-tipped' eagles and ravens in feasting on the slain, or at least so they did in Anglo-Saxon poetry.

Legislation protected game animals in forests, but although the wolf was also hunted, it was never designated as game. It was always the enemy, always vermin. It killed and devoured the king's deer, and the husbandman's calves and sheep, and the cottar's cow. It forced the landowner to keep guard dogs and pay the wages of herdsmen and huntsmen. By the thirteenth century, it seems, there was a concerted effort to rid England of wolves. A bounty of five shillings – roughly £500 in today's money, and then, equivalent to 25 days' pay for a skilled worker – was offered for the head of one presumably elusive and annoying wolf. With that, a poor man could buy a milk cow, or a quarter (about 500 pounds) of wheat, enough to keep the family in bread all winter. The inference is that wolves were already becoming scarce and shy and hard to track down.

Later that century, King Edward I institutionalised extermination by organising the lupine cleansing of forests all along the wild Welsh borderlands. Those who held land from the king were obliged to 'defend the land from enemies and wolves' as part of their feudal duty.

Wolves were hunted in winter-time – January was *wolfmonat*, the wolf-month, the start of the hunting season. It was partly because winter pelts were thicker and more valuable, but also because wolves raise their cubs early in the year, and so were at their most vulnerable at that time. Once extermination became the aim, perhaps more wolves were trapped in baited pits than were hunted with dogs. The odds could be further levelled by smearing pitch to spoil the wolf's scent trail, or by using fladry, a kind of bunting, to contain the frightened animals within a chosen area, or simply drive them towards 'blinds' where they could be shot. Alternatively, a skilled huntsman might track prints in the snow or mud all the way to the den and spear the cubs. However they did it, the English and their wolves seem to have parted company by the end of the Middle Ages. Henry VIII, for instance, who was a keen huntsman, seems never to have encountered a wolf.

In the wilds of Scotland, wolves lasted longer, in fact another couple of centuries longer thanks to the difficulties of controlling wide-ranging animals in the Highlands – and also to the ready availability of fresh wolf-food in the form of deer. With the firm intention of getting rid of them, wolf hunting was legalised, and indeed made compulsory. In 1427, the king enacted a law requiring a minimum of three wolf hunts during the cubbing season. Mary Queen of Scots was entertained to a famous wolf hunt in the Forest of Atholl in 1563, in which the royal party managed to kill five wolves, among a sizable pile of dead deer, foxes, martens and wild cats. This seems to have been among the last formal wolf hunts. Perhaps the demise of the wolf was linked to the development of firearms, just as the extinction of the great bustard in Britain coincided with the first reasonably accurate long-range rifles.

Hatred of the wolf was reinforced by age-old folk tales: Little Red Riding Hood and company. Like, say, poisonous mushrooms or venomous spiders, fear exaggerated the danger. In fact, there does not seem to be a single well-documented case of a wolf killing a human being in Britain – although such knowledge might not necessarily calm the nerves of a traveller on a lonely moor at night. There were also concerns about wolves disturbing the resting places of the recently dead. Lest the wolf should dig up and devour the corpse of a loved one, in some coastal parts of Scotland graveyards were consecrated on offshore islands beyond the reach of the wolf. There was no talk by then of the nobility of wolves, no more wolf-names or wolf-places. They were routinely described as 'rapacious', 'ravaging' and generally 'noisome'. Wolves were explicitly enemies of mankind and hence, by another small stretch, ascribed as being agents of the devil.

In those circumstances even a lone wolf became one wolf too many. In one of the last records of a Scottish wolf, in 1621, a laird noted in his diary the astonishing sum of six pounds, thirteen shillings and fourpence (equivalent to £1,700 today) paid to one Thomas Gordon for his trouble in tracking down and killing one notorious animal. Shortly after that, the wolf fades into myth. Did Sir Ewen Cameron of Locheil really shoot a lone wolf at the Pass of Killiekrankie in 1680? Did MacQueen of Findhorn actually kill a 'black wolf' at Ballachrochin, near Inverness, in 1743, or was it actually a hybrid, like White Fang in Jack London's novel, or even just a big feral dog? A social animal like a wolf surely needs to be present in numbers or not at all; the persistent survival of a few lone wolves seems unlikely. After making exhaustive enquiries, the naturalist Thomas Pennant, writing in 1769, was satisfied that there were no more wolves in Scotland.

Sir Ewen's alleged wolf of Killiecrankie was stuffed and displayed in a glazed cabinet, arranged in a crouched position, licking its chops. It was subsequently bought by a curio collector, Sir Ashton Lever, who got an artist to paint its portrait. But for its grey mane you could mistake it for an Alsatian. You could easily imagine it curled up by the hearth chewing a bone. Its portrait was painted again some years later by the naturalist Edward Donovan. His is a cruder and fiercer image, an alert and wide-eyed animal showing lots of sharp teeth. Perhaps that is how Donovan perceived a wolf should look, just as so many old pictures of wild cats show them snarling. Maybe he thought he could charge more for a properly fierce wolf. The stuffed Last Wolf is now lost. All that remains are the two portraits, which we could call Smiley and Snarly.

In Ireland, too, the wolf was seen as the enemy, and extermination the aim. As in Britain, the least populated places, especially in the far west, were the wolf's redoubts, but odd animals were noted elsewhere, even within a day's ride from Dublin. Wolf-control measures, with offered bounties, were passed in 1614, and again in 1652 and 1653, in the latter years as part of Cromwell's conquest. The rewards for a wolf's head were substantial enough to attract bounty-hunters. At one point the reward reached six pounds (£640 today) for an adult female, the equivalent of 66 day's wages for a skilled tradesman, or five pounds (£535) for an adult male, and 10 shillings (£50) for each dead cub. By then, it seems, the wolf was not so much a danger as endangered. Extermination had succeeded in Ulster by the 1690s, and over the next fifty years it was wiped out in the wilder west of Ireland too. The last known Irish wolf was hunted down in County Carlow in 1786 after it had killed a sheep. And that, at last, marked the end of living with fierce wild beasts.

The extermination of the British and Irish wolf is charted in faded lines in dusty ledgers accessible only to scholars: a fragmented record of bounties paid, of laws enacted, of animals bludgeoned, shot, or skinned. There are no surviving pelts, and not even many bones, the twelfth-century skeleton of the Helsfell Wolf in Kendal Museum being one of the few. And yet the remembrance of our last fierce animal lingers on – the echo of wolf howl sounding down the ages. Today we can afford to be nostalgic about the wolf, perhaps to regret its disappearance, and even to see its loss as an ecological crime. But what is it that we feel we have lost? An animal for which, as a nation of dog-lovers, we feel a bond? Or a symbol: an evocation of the wild? Or even the embodiment of a sense of atavism, of a half-wish to leave our over-tamed countryside and step into a more fearful beyond? Whatever it is, and however it is felt, the wolf continues to have a presence in our ecologically-aware consciousness. Its restoration is the ultimate aim of re-wilding. Perhaps we feel a sense of injustice. Do we feel we owe it to the wolf to restore it to the landscape – just as so many other European countries have done?

We certainly live with the consequences of the lack of wolves and other big carnivores. Without predators other than human ones, deer now chew away the regrowth in woods, and help the sheep to reduce the Highlands to an over-grazed, species-poor emptiness. Without them red deer routinely starve to death. Restored to Britain from continental distant cousins, wolves and lynx would restore the balance of nature. Both animals have returned to parts of Europe where they are more-or-less tolerated (though not always; the first wolf seen in Belgium for at least a century, in 2018, disappeared shortly afterwards, and was probably shot).

Even though our memories of living with wolves are more distant than anywhere in Europe, it seems unlikely that we will ever again walk with wolves unless the latter are secured inside tall, electrified fences. We can probably forget about wolf packs howling free on the North Downs, or of eyes in the undergrowth watching the golfers on Cotswold commons. Ours will be a slow, cautious ecosystem-building, a step-by-step approach in which health-and-safety – ours that is – will count for more than ecology. No Yellowstone or Yosemite here, but perhaps some sort of game park, with tagged animals, probably somewhere in the northern Highlands where they need the tourists.

Perhaps restoration isn't really the point. The wolf, in Britain, represents a dream of the wild. As George Monbiot says it, 'I want to see wolves reintroduced because wolves are fascinating . . . they seem to me like the shadow that fleets between [heartbeats], because they are the necessary monsters of the mind, inhabitants of the more passionate world against which we have locked our doors'. For the time being, the British wolf remains a dream. Perhaps, if we are honest, most of us would be happy enough dipping our toes into a surrogate wilderness. We have lost our taste for danger.

Hell on the Yangtze: the baiji

Most of the large animals of the world live within only a shadow of their former range. The lion of North Africa and the Middle East, for instance – the beast of the Bible and the Greek myths – has all but vanished, and even the great whales have departed from parts of the oceans. But the biggest losses of megafauna are among a group that we seldom think about: the great beasts of freshwater.

Across the world there are, perhaps surprisingly, around two hundred freshwater animals, reptiles and fish weighing over 30 kilograms – that is, as big as us or bigger. They include freshwater seals, porpoises and dolphins, and large fish such as the beluga sturgeon, the Chinese paddlefish and the Amazonian arapaima. And pretty well all of them have crashed in numbers in recent years. A recent study (He et al. 2020) concluded that freshwater large life declined globally by 88 per cent between 1970 and 2012 (and let's add a few more percentages for continued losses since then). Half of them are now considered to be threatened, and thirty-four are critically threatened. The other half aren't necessarily safe either, for many species are 'data deficient', meaning that we don't know enough about them to make an assessment. Given the state of the places they swim, which include the shrinking Mekong and the polluted Yangtze, the authors of the paper weren't very cheerful about their prospects either.

At least one species of the freshwater megafauna has probably gone for good. This unlucky beast was the baiji, or Yangtze river dolphin of China. River dolphins are an ancient group that have taken to life in the great broad rivers of the world. There are just three species (or five if you count two South American segregates): the Asian, Amazon and Yangtze dolphins. Compared with their sea-going relatives, they have bendier necks, narrow, beak-like jaws, small eyes and a characteristic bump on their foreheads, known from its shape as a melon. Their paired front fins are longer than those of ocean dolphins and have become paddles, a necessary adaptation to the swirling currents of the river. Since visibility is often poor in these muddy waters, the dolphins rely on sonar or echo location to locate their fishy prey. The melon on their head generates high frequency clicks which find an echo in any fish nearby, received by way of sensors in the dolphin's throat and passed to its inner ear (a way of hunting that is hard for we humans, who hunt by eye, to imagine, though perhaps a bat would understand). Quite apart from their well-tuned senses, these dolphins are by far the most intelligent life in the river.

The Yangtze dolphin is the rarest of the three (or five) for it is restricted to a single river, the third longest in the world and the greatest in Asia. The Yangtze runs 6,000 kilometres, from its source in the Tibetan plateau to the sea via a delta and the port of Shanghai. One third of the population of China lives within its enormous catchment. The river is, or was, home to its very own species of alligator, sturgeon, porpoise and paddlefish. But the most famous of the river animals was the Yangtze dolphin, a being traditionally regarded by river folk with reverence. They called it *baiji*, meaning 'white fin' (for the little triangular fin on its back, visible when it surfaces, is white). Legend had it that the baiji was once human. She was an orphan princess who dived into the river, to escape her evil stepfather who was taking her across the river to sell her. Being good, she was transformed into a dolphin. The stepfather was also washed overboard and, being less good, was turned into a porpoise. The baiji became a kind of river goddess. It looked after the river. It became a kind of symbol of ecological harmony.

There were still sustainable numbers of baijis in the Yangtze within living memory. But the Chinese economic miracle did for them, as it also did for the three-metre-long Chinese paddlefish, and may yet do for the alligator, the porpoise and the sturgeon. Threats to the baiji came from multiple directions and their collected weight must have turned its existence into a prolonged agony. As the Chinese economy boomed, the traditional fishing boats of the river were replaced by container ships, car ferries, tankers and trawlers, and colossal freighters carrying Brazilian iron ore to the steelworks. As the river traffic increased, so its continuous roar scrambled and disabled the dolphins' sonar. They were quite literally deafened by the scream of engines. At the same time the trawlers with their huge dragnets beat the dolphins to the fish and scored their flesh with the bite of their propellers. And then Chinese engineers built the Three Gorges Dam, which, among other things, increased the sediment load of the river so much that even the dolphins were blinded by the murk. And the water itself became poisoned by a witches' brew of industrial effluent, chemical fertiliser and sewage. It all happened within a generation. Under the political philosophy of Chairman Mao, nature was the enemy. Mao made war on nature. The traditional veneration of the baiji was officially discouraged, indeed denounced, perhaps by the vigorous means that the Chinese Communist Party had at its disposal. It followed that no particular efforts were made to save the baiji, or at least not until it was too late.

Dolphins are sentient animals with acute senses, and the baiji will have suffered before they died. If we cannot quite imagine how they lived, we can surely sense how they might have perished, choked, deafened, blinded, starving and terrified. There was no escape. Maybe they searched desperately for calm, clean water and found none. They spent more and more energy trying to locate fish. They might even have had trouble finding one another in the poisoned murk and, when they did, it was likely to be an encounter of exhausted, sick animals, perhaps bleeding from wounds and with their gobbets of dead flesh nibbled by the river's small fry (the minnow's revenge!). How many females, which mated only once every two or three years, were in any condition to give birth and bring up a calf? What chance had they got? And, you have to wonder, did the animals experience terror? Did they, as they smelt the stench of toxic chemicals, as their world grew dark and the natural sounds of the water turned into industrial bedlam, experience a premonition of their likely fate? As Rob Newman noted in his radio broadcasts, *The Extinction Tapes*, 'We're always told that only humans have a sense of their own mortality. That we are the only animals that know we are going to die. But if other animals sense danger, then they surely sense danger of death?'

The figures published by the IUCN speak for themselves. In 1950, there were perhaps 6,000 baiji in the Yangtze. By 1970, there were only a few hundred left. By the 1980s, a more thorough survey found just 400. In 1997, after a 'full-fledged search', they found just 13. In 2006, another intensive search over the full length of the river, using two research vessels equipped with state-of-the-art acoustic sonar, failed to find any baiji at all. 'It just happened so quickly.' The baiji was declared 'functionally extinct', the caveat being necessary because, even if a few still existed in some back-water somewhere, their numbers would now be too few to sustain the species. The last photographed baiji was in 2002. The last verified sighting of one was in September 2004. Latterly the Chinese government did make belated attempts to save it. A conservation plan was drawn up with 'Baiji Reserves' in various, relatively quiet corners of the river.

Plainly the plan failed, and it was regarded internation-ally as 'not a serious effort'. There were no captive baijis available for an attempted reintroduction. River dolphins captured for dolphinariums and zoos in any case rarely survive longer than a few months. Of hundreds captured, mostly of Amazon dolphins, only four were still alive in 2015 (and there is now just one, in Peru). And so the Yangtze river dolphin became just a memory on troubled waters, another stain on our environmental record.

The awful fate of the baiji illustrates an easily overlooked aspect of extinction: that the end of a species is more than a statistic or a name on a list. It is about the crushing out of unique life in a cauldron of pain, suffering and terror. The existence of the last individuals of a sentient species like the baiji would be darkened by the inevitability of its fate, in the complete void that awaits them once their (and our) environment has been systematically trashed. It is easier to empathise with a dolphin because they are such attractive animals, beautiful, intelligent and, in captivity, so friendly, forgiving and eager to please. Perhaps it is harder to feel the same way for the dying vultures of India, poisoned by veterinary drugs, and harder still for the dung beetles of Europe, dying out for similar reasons: death by toxic dung.

Of course nature will have a way of making us regret what we have done to the vultures and the beetles when we find that animal corpses now lie stinking until they rot, contaminating the water supply and spreading disease, and when dung is no longer recycled to enrich the soil. But beyond that there is also the small matter of how we might feel about ourselves, that we do such things. Is it manifest destiny? To improve our lives, does it follow that other species must be obliterated? Is it possible, even, to think of the extinction of the baiji as, on balance, a positive thing, as a milestone (as opposed to a tombstone) of progress, of the annihilation of an animal irrelevant to our concerns, one that just happened to be in the way?

Chapter 6

Afterlives

To die, to sleep –
To sleep, perchance to dream.
> William Shakespeare, *Hamlet*, Act III, scene i.

I will not say
That he is dead. – He is just away.
> *Away*, James Whitcomb Riley (1913).

Let's cheer up. Let's get upbeat. Perceived extinction isn't necessarily the end of the story. Sometimes species that scientists believed extinct turn out to be very much alive, especially secretive or obscure species, or those lurking in remote places, such as the deep sea. The classic example is the coelacanth, a fish believed to have died out millions of years ago, which was found alive and well in 1938, in the ocean depths, presumably as unaware of humankind as we were of it. The world is still full of hiding places.

The reason why the conservation industry is often reluctant to declare a species extinct is that it is almost impossible to verify the loss unless, that is, its habitat has been completely destroyed. It seems more prudent to call them endangered. Beautiful and exotic plants can and do come back from the dead for they may survive un-noticed as buried seed for decades or even longer. Even among such well-watched species as birds, nature has the capacity to surprise us. The memorably-named black-browed babbler of Borneo was found alive and well in 2020 after having been written off as extinct for more than a century. Such species, rising from the grave as it were, are called Lazarus species, after the man in the New Testament who died and then came back to life. I mention some, perhaps unexpected, examples in this chapter.

There is a different kind of afterlife: animals and birds which have become so famous that even death cannot put an end to their charisma. They live on in a kind of radiance, in our imaginations, and in our products and images, including films, more famous now than they ever were in life. At least one animal that never existed shares that kind of cultural afterlife glow: the dragon. I claim the dragon as an extinct 'species' in the sense that people once believed it existed and made notes on its imagined natural history. It was, in a sense, not only a real animal but a very important one. This chapter is devoted to the different forms such afterlives can take, as exemplified by the extinct dodo, the long extinct Tyrannosaurus or *T. rex*, the once extinct but later rescued British military orchid and, of course, the non-existent dragon.

Dead as a dodo

Everyone knows the dodo. It is one of the most famous birds in the world. We all know what it looks like: an improbable fowl, as big as a turkey, with a body as full and round as a plum pudding and puny, useless wings – but, as if by way of compensation, a big heavy beak with a useful looking hook at the far end. As the American humorist Will Cuppy wrote, the hopeless looking dodo seems to have been invented for the sole purpose of becoming extinct.

It lived only on the island of Mauritius, among other bizarre beasts: a giant tortoise, a red flightless rail with a long, curved bill and a near-flightless parrot with another outsized beak. Partly perhaps because of its funny name, but also because of its ridiculous appearance, the dodo has become a byword for evolutionary obsolescence: 'as dead as a dodo' means very dead indeed, about as dead as a doornail. By the same token, the poor bird has also become associated with blundering stupidity.

Back in the nineties, the Tesco supermarket chain ran a series of short adverts featuring a hapless dodo called Derek. Every time Derek hit the kitchen he managed to 'extinct' himself. All set to carve the Tesco turkey, he slices himself up instead. Heating the Christmas pudding, he catches fire and burns to ash. Biting a frozen prawn he turns into ice and falls to pieces with a clatter. The tag line is: 'Buy now, they won't be around forever'. As a fall guy the dodo is always there for a laugh.

The dodo is an example of a good evolutionary idea that turned into a bad one. The food supply for a big hungry pigeon on a relatively small, isolated island may be limited. Buds, berries and seed are seasonal, and so there may be times when there is nothing much to eat. One way out is to grow large and plump, so that your body fat will tide you over until the good times return. The downside to this strategy is that you may become too plump and heavy to fly. But that is no great problem so long as there are no predators or competitors around, or at least none that you haven't grown up with. Unfortunately this logical development lays you open to disaster once more aggressive, cosmopolitan species discover your island refuge. And so it turned out.

Flightless birds on isolated islands are usually among the first species to go once the ships arrive, as I've already discussed. The time when human beings and the dodo shared the same living space was short: no more than a century, and probably less. The first European visitors to set foot on Mauritius arrived in 1598, in a squadron of Dutch ships under Admiral Van Warwyck. The island soon became a useful place for ships to anchor for water and fresh food, and for repairs, and was eventually settled permanently by Dutch immigrants from 1638. An engraving made in the early 1600s depicts a temporary settlement in which everyone is busy. The ship's company are netting for fish, sawing wood and sealing barrels, cooking a meal, probably fish or turtle, or are grouped around a preacher, listening to his sermon. On the margins and seemingly ignored are various strange birds and beasts, perhaps attracted there by the commotion but now moving away. Among them is a crudely drawn but unmistakable dodo. Perhaps, with the innocence of a bird that had never before seen a human being, it had come to peer at this strange new two-legged animal, but then thought better of it. Later, perhaps, someone will take his gun into the bush and they will get a chance to see what roast dodo tastes like. The last documentary record of a dodo was in 1662, only twenty-four years after the first settlers came to the island: the bird was almost certainly extinct by 1680.

As to whether the dodo was tasty: it was really a giant pigeon and so it ought to have been. But dodo flesh was apparently disappointing, tough and greasy, hence its Dutch nickname, *walghvogel* or 'disgusting bird'. One witness noted dispassionately that, when caught, the terrified bird would let out a scream. Its cry attracted other dodos who would rush in to help, and so they were knocked on the head too. What particularly interested the settlers was that, when cut open, dodos were often found to have swallowed stones, some as big as a fist or a hen's egg.

The various names given to the living dodo must have reflected what people thought about it. Dodo itself seems to come from the Portuguese word, *duedo*, meaning 'fool' (sixteenth-century Portuguese sailors stopped off at Mauritius for food and water). Perhaps the bird's foolishness lay in its tameness and curiosity, perhaps it moved rather oddly or maybe it just looked badly put together and silly. As puns on its name, the Dutch jokily called it *dod-aars* or 'fat-arse', or *dodoor*, meaning 'lazy', an idle bird. Other names included *griffeendt* (from its formidable griffon-like bill?), Kermis goose, Walck bird or, still more mysteriously, Birds of Nazareth. Perhaps it is just coincidence that dodo is an anagram of *dood*, Dutch for 'dead'.

The best description of a living dodo is in a 1634 book of travels by Sir Thomas Herbert, an English courtier and historian. The bird, he noted, was round and fat, weighing at least fifty pounds. It was clothed in downy feathers, with three large plumes on its tail. What struck him most was the dodo's expressive eyes, as bright as diamonds, 'round and rowling'. It had a strangely sad expression, as if 'sensible of Nature's injurie in framing so great a body' with 'wings so small and impotent'. But in the most lifelike image of a dodo, a watercolour of the head and neck by one Cornelius Saftleven (1638), the dodo looks anything but sad; on the contrary it seems intent and eager-looking, a bit like a favourite pony expecting a sugar lump (Fuller 2002).

Given that human beings shared the island with the dodo for several decades, it is striking how little we really know about it. We have no idea of its numbers, for instance, or where it lived (the possibility that the dodo was a coastal bird is conjectural); we don't know what it ate, although it is presumed to be a fruit-eater; we don't know how long it lived, and not much idea of what it sounded like either, although one witness compared its cries with that of a gosling. We don't know whether the male and female birds were differently coloured. For that matter, we don't even know for certain what its colour was, for not a single feather survives. Some pictures suggest that the bird was dark, perhaps dark grey or dark brown, with paler skin on the naked flesh around its eyes and bill, and with black legs and yellow feet. It had sharp claws, noted Thomas Herbert, and thick legs 'suited to her body', and with an appetite 'strong and greedy'.

Most pictures show the dodo as round and fat, almost impossibly plump, but it seems they may be wrong. Recently the Natural History Museum made a 3-D model of the bird, using laser surface scanning technology, which produced a much slimmer, more plausible bird with a longer neck and, interestingly, a good brain with acute senses for sniffing out food on the ground. Modern weight estimates are smaller than Thomas Herbert's guess of fifty pounds. Around 10 to 22 kilograms is more likely, still a heavy bird but one lighter on its feet than the usual image. Quite possibly the pictures of the canonical overweight dodo were based not on wild birds but either on fattened-up captives or over-stuffed dead ones.

Outside of Mauritius, the dodo was not well known. A few live birds were exported to Europe and the East but they did not survive long. A living dodo was on show for a while in London but failed to attract much interest. Few relics survive. There are two skulls but the only recorded complete, stuffed dodo, at the Ashmolean in Oxford, was judged to be of so little importance that the remains were thrown on a bonfire. Only the head and right foot were saved – the last soft tissue of the dodo left on earth. It was a subsequent fake dodo in Oxford's Natural History Museum that caught the attention of Charles Dodgson, alias Lewis Carroll, and his young friend, Alice Liddell. He incorporated it into *Alice's Adventures in Wonderland* as an in-joke. Dodgson had a stutter, and so pronounced his name Do-do-Dodgson, hence his nickname, 'Dodo'. His illustrator, John Tenniel, based his own picture on the Museum's dodo, and so created the image we all know today, the plump, kind, even fatherly dodo leaning on a cane and handing Alice a comfit with a mysterious hand emerging from sleeve-like wings.

We don't know how or why the dodo became extinct, only that it did. Perhaps the clearance of the island's natural forests, so rich in fruit trees, condemned the dodo to a lingering demise. With it disappeared all the endemic tortoises and parrots and other flightless birds that shared its isolation. Some believe that it was not humankind which killed them off so much as the beasts that accompanied him in the ships and which then escaped: monkeys, cats, pigs, rats, mice. The tortoises and flightless birds had never before been in contact with such animals. They were presumably defenceless against the hungry, cosmopolitan outsiders who eagerly helped themselves to their eggs and hatchlings. Mass extinction tends to follow such invasions, as many an ocean island can attest.

The post-mortem progression from the last dead dodo into a popular icon of lost life came about in the first instance through Lewis Carroll and John Tenniel, but in a wider sense it was also because, even in a world full of strange birds, there was never anything else quite like a dodo. Rightly or (more probably) wrongly, it has become the image of an animal that was too gross to live: heavy, round, fat and greedy, it had tottered about in its island Eden, free of enemies, apart from slow-moving land crabs, until the outside world caught up with it. Yet, unlike most lost birds, the dodo is not forgotten.

In Mauritius today, there are dodos everywhere, on coins, banknotes and knick-knacks, and even supporting the island nation's coat-of-arms. They abound in every shop, on matchboxes, stamps, crockery, tea towels and tins. There are dodo drinks and dodo burgers and dodo chocolates known as Whoopsies. Some of these posthumous dodos have become cheerful cartoon characters wearing baseball hats and striped vests. My favourite dodo product is a special mat that *deadens* the sound: a 'dodo mat'. And so the dodo's iconic status has lent it a kind of immortality, a fool's progress from the natural world to our own. In death it became one of the most famous birds that ever lived.

A different kind of immortality came to the dodo's less-known relative on the small, isolated island of Rodrigues, 600 kilometres further out in the Indian Ocean. The solitaire, or solitary, was, like the dodo, a giant flightless pigeon, but one shaped more like a goose with a long neck and a more conventionally pigeon-like head. Like the dodo, it was 'sad of face', and when caught was said to shed tears.

Apparently, the solitaries fought one another using their little stubby wings as clubs. Unlike the dodo, they tasted quite nice. Only one person ever described the bird and drew its likeness, and all other images are based on that single sketch. Like the dodo, the solitaire did not get on with humankind; it was soon gone, leaving only a scatter of bones, and not a feather nor a scrap of flesh survives. With it also vanished the Rodrigues pigeon, the Rodrigues starling and the Rodrigues owl. It was obviously bad luck to be a species with Rodrigues in your name.

In 1761, the astronomer Alexandre Guy Pingré made a voyage to the island to observe the transit of Venus across the sun (a rare event and then important because it enabled the diameter of the sun to be calculated). He was informed that a large flightless bird lived on the island, and he wanted to see it for himself. It seems, though, that no one could find one for him. A friend of the Abbé, the astronomer Pierre le Monnier, decided to honour the visit by raising the lost bird to the heavens. To that end, he borrowed a few faint stars from existing constellations and named them *Solitarius*, the solitaire. It was a stellar apotheosis for the solitaire! There it floated in a mist of stars, perched on the tail of the Hydra and in some danger of being clouted by the scales of Libra. Unfortunately, the starry solitaire did not last long. Images on later star charts suggest that it was soon transformed into a different kind of solitaire named *Turdus Solitarius*, the blue rock thrush of the Philippines. Later still, that bird was transformed into an owl, *Noctua*, before being forgotten altogether. Interestingly, on a British star map published when we were at war with France, they substituted a mockingbird. *Solitarius*, it seems, was an unusually versatile constellation.

While the solitaire is almost forgotten, its fellow monster

pigeon lives on as an icon of extinction, as a suitably ridiculous example of sudden vulnerability. That the dodo was wiped out so casually, and so quickly, offers a lesson in island ecology, while its continued existence as an icon suggests an interesting unwillingness to let it go. We know a good extinct species when we see one. Perhaps the dodo dead was always destined to be more famous, in a sense more useful, than the living bird, with its unknown lifestyle and its never heard song.

Shall we leave the last word to Will Cuppy? 'Most of us feel that we could never become extinct. The Dodo felt that way, too'.

The many faces of *T. rex*

I keep meeting that mighty and long-lost animal *T. rex*. He seems to be everywhere, not just in museums but in films, in products, in miniature model form and even in advertising. He is so popular he gets the generic 'T' as if it were an initial like Tony or Tom, instead of *Tyrannosaurus*. But he is also 'rex', the king, even though a full half of him must have been female and his best-known skeleton is called Sue. There he stands in his pomp and glory, ruler of all he surveyed, 67 million years ago, or at least the bit of the earth called Laramidia in what is now western North America. And, as we all know, he was still lording it when something like a comet appeared in the summer sky and shortly afterwards hit the earth square in the Gulf of Mexico burning the tyrant king and all his subjects to a crisp in the last remaining seconds of the Cretaceous.

T. rex entered my life early, in Walt Disney's film *Fantasia*. The tyrant king was given a walk-on part in which a lush vista of placid monsters quietly munching water plants was suddenly transformed by him into a scene of terror. To the pounding chords and thudding timps of Stravinsky's *Rite of Spring*, he came stomping into frame, red-eyed and ferocious, and implacably hungry. All the other cartoon dinosaurs fled for their lives, but a mountainous stegosaur was too slow and was forced to stand and fight. *T. rex* made short work of it, and the death throes of the expiring stego were matched by the music, chord for chord. You could imagine every small boy in the cinema whispering to their parent, 'Dad, you know *T. rex* and Stegosaurus didn't live at the same time? It should have been the Triceratops. Oh, and Dad, did you spot the *Parasaurolophus*?'

Tyrannosaurus has been stomping along like that since the 1920s. The Natural History Museum used to sell a range of state-of-the-art plastic dinosaurs and, of course, they included *T. rex*, even though he lived on the far side of the world from the museum. He was modelled in crimson, with a very long tail and a curiously flattened head. Given that the real beast was about forty feet long, taller than a double decker bus and weighed more than a full-grown bull elephant, it was hard to imagine how this version of him could move at all, dragging his heavy tail behind like a ball and chain. The stop-motion animation of films like *The Lost World* and *King Kong* didn't help, for their *T. rex*es were obliged to jerk along like ducks. Our dinosaur might have looked pretty impressive standing still, and inspiring the likes of Godzilla, but the pre-1970 animal was a puzzling beast. The only thing about him that made sense was that he was extinct. Unable to catch his dinner, he had presumably pegged out from a sense of ennui and general exasperation.

Things began to change after a thorough study of a new kind of dinosaur called *Deinonychus* (the 'Velociraptor' of *Jurassic Park*). This animal, whose name means 'terrible claw', had the same basic build as *T. rex*, that is, two-legged and long-tailed, but in its case the tail was stiffened by tendons and was clearly used as a balancing organ, just as a cheetah relies on its tail to perform high-speed swivels and turns. *Deinonychus* was evidently an active predator, able to outrun and then rip at its prey with those terrible claws. There was something distinctly bird-like about it, and modern restorations of *Deinonychus* now clothe it in feathers: a giant, flesh-tearing roadrunner, as terrifying, in its way, as *T. rex* himself.

Deinonychus encouraged scientists to imagine predatory dinosaurs in a different way. They junked the old static models of *T. rex* and transformed them into the much more agile *T. rex* of *Jurassic Park* (1993). They tilted his body forward by the hips until it was more-or-less horizontal, thus lifting his now shortened tail up into the air. Suddenly *T. rex* could run – though maybe not as fast as he did in *Jurassic Park*. Recent studies of dinosaur mechanics suggest that any seven-tonne dinosaur attempting to overtake a Land Rover would quickly collapse in a heap with both legs shattered. But all the same, we now had a more plausible beast that could at least out-pace its likely prey, while also proving pretty adept at the consequent ripping and tearing. Studies of the *T. rex* jaw suggest he had a bite-force more than three times that of a lion: he was a biter *extraordinaire*, a cruncher of bones. Science transformed *T. rex* from a cartoon beast into a proper predator – one that we can imagine and relate to, and also to thank God there is nothing like him left on earth.

Since then there have been a few more refinements. The digital miracles of *Jurassic Park* continued with the BBC's 1999 *Walking with Dinosaurs* series, and similar reconstructions on the American-based Discovery channel. Their dinosaurs moved and behaved plausibly, and mostly within the scientific evidence. But the *T. rex* which appeared towards the end of *Walking with Dinosaurs* was one of the less successful actors. The team had experienced trouble modelling the head, which they interpreted as boxy and toad-like, with piggy eyes buried under projecting 'hornlets' and a bumpy ridge extending down his snout. That didn't look right, somehow. Nor could they convey how *T. rex* fed, for the close-up, manually operated rubber heads shown picking at their food were among the least convincing aspects of the show. There were also signs of our own world intruding into the Cretaceous. The BBC's *Tyrannosaurus* was a caring dinosaur, a good mother which built a nest of compost for her eggs and looked after the babies until they could fend for themselves. It's possible, just about. But then I read a book whose author claimed that the intelligence of *T. rex* had been grossly underestimated. It might even have been as bright as a chimpanzee! (It was something to do with 'encephalisation quotients'). Yes, even scientists are apt to get carried away by wish fulfilment. Since the evidence is slim and often ambiguous, we are left free to create the dinosaur we want: unexpectedly clever, caring, and perhaps even a bit of a feminist. In children's books and cartoons, *T. rex* can look smiley and quite friendly, a much misunderstood creature, perhaps. But in years past, he was more often presented as a kind of outlaw from the Wild West, always angry, always ferocious, forever hungry, and with the brains of a particularly dim-witted crocodile.

We constantly reinvent him to fit the concept of *T. rex* that suits us best.

The fully up-to-date *T. rex* – at the time of writing anyway – has a fuller, chubbier head than the BBC beast, with more flesh on his bones and fewer bits and bumps. Some would even give him lips. Relatives of *T. rex* have been found with feathery down preserved on their bones, and so it is possible that he too was a feathered dinosaur, at least in his youth. Most representations now give the full-grown animal at least a few feathers, and some cover him with down from head to foot so that he looks like a giant emu. Feathers bring in the question of colour. The *T. rex* of *Jurassic Park* and *Walking with Dinosaurs* was a nondescript grey with vague blotches and a paler counter-shaded underside, true to modern perceptions that big animals are mostly dull-coloured. Feathers, though, suggest brightness, at least in patches. It is anyone's guess what colour *T. rex* actually was. But between them the hundred-odd *T. rex* toys on the market today cover practically every hue in the paintbox. One of the newest has sky blue feathers running from head to tail, plus a tuft of bright red plumes between his round, staring eyes like a punk haircut. He looks like a giant parrot.

Whatever he might have looked like in the flesh, *Tyrannosaurus rex* is easily the most famous of all dinosaurs. It is because of his popularity that museums pay so much for a skeleton ('Sue' cost the Chicago Field Museum $7 million). It is also why collectors go to such lengths to seek his remains, and why scientists have been able to examine them so minutely. In fact *T. rex* was a mega-star almost from day one. Barnum Brown, the museum curator who excavated the first recognisable bones in 1902, always referred to *T. rex* as 'my favourite child'. Henry Fairfield Osborn, who described the species in 1905, called him 'the *ne plus ultra* of the evolution of the carnivorous dinosaur' and one 'entitled to the royal and high-sounding group name' of *Tyrannosaurus*, the 'tyrant lizard'. *The New York Times* hailed him as 'the most formidable fighting animal of which there is any record whatever . . . the king of kings in the domain of animal life . . . the absolute warlord of the earth'. A little later, as the first museum skeleton was going up, the same paper acclaimed him as 'the prize fighter of antiquity'. Big fierce animals have always been admired, and this was the biggest and (presumably) baddest land carnivore that has ever lived: fighter, sovereign, warlord. Today, he has some equally hefty rivals in other continents – *Mapusaurus*, *Giganotosaurus*, *Carcharodontosaurus* – but they lack the charisma of the original. *Tyrannosaurus* rules. No one is going to get very excited by *Carcharodontosaurus*, even when they are able to pronounce it.

Animators have become so skilled in creating plausible *T. rex*es that it is almost as though the great beast has come back to life again. We can imagine him as readily as a tiger. That being so, it does invite the intriguing question of just how dangerous *T. rex* would be, assuming that he could be magically teleported from the Cretaceous to the present day. Many of us will have enjoyed the full *T. Rex* Experience at a museum, loitering within a few inches of those plastic-coated, digitally controlled open jaws. The Natural History Museum has a life-size automaton which presides over a 'family-friendly eatery' they call the *T. Rex* Grill. When I was last there, he obligingly lowered his head to mine, revealing a cage of banana-sized teeth. The Oxford Natural History Museum has a less showy cast of a skull, similarly toothy and gaping, into which I once pretended to stuff my young companion, a keen dinosaur fan. He squealed with delight. As this demonstration indicated, *T. rex* could easily have swallowed a human whole, along with his hat, briefcase and brolly. But, you wonder, would he have bothered to do so? There are some super large horseflies that nevertheless seldom bite humans simply because we are not worth the trouble. They have evolved to bite bigger, more impressive four-legged animals and fail to recognise us as prey. Quite possibly *Tyrannosaurus rex* wouldn't either. Dinner for him would be another dinosaur. Conceivably the danger he represented to those scientists in *Jurassic Park* would come not from his jaws so much as his great lumbering feet. The one they really needed to watch out for was *Deinonychus*. We are about the right size for those.

Today's children know *T. rex* as well as they know giraffes and elephants. Better, probably. This long extinct animal has attained a kind of visual equality with living ones. The further we are removed from the real thing, the greater is the force imagination brings to our perceptions. In that sense, extinction, for animals like *T. rex*, is not the end. He lives on as an icon, a brand, a reimagined beast that we can relate to and bring into our lives if we wish. He might be on our T-shirt, turned into a lampstand, dangled from a keyring or snarling on someone's arm as a hard guy tattoo. Far from being frightened of him, young children can use him as a push toy, or ask for Christmas that strutting, roaring *T. rex* made in China. A resin cast of his skull can make a popular conversation piece. I have even spotted a *T. rex* toilet-holder, perhaps the ultimate degradation of the tyrant king (the idea is that his great teeth keep the roll in place). When one of the most famous glam rock groups took his name, no explanation was necessary (and none was given either). 'T. rex-ness' is a quality as well as a biological reality. Leaving aside the purely whacky, it brings with it a sense of power, of aggression, or even of irony (like the tattoo design that has him wearing a top hat). One dinosaur authority, Thomas Holtz of the University of Maryland, expressed it thus: 'T. rex appeals to both the high-minded and visceral parts of ourselves. The poetry of the species has a persistent draw' (Black 2018). More than any other extinct beast, he has been drawn into our world and remade.

T. rex has not gone forever. T. rex lives! He always will, in one shape or another, as tyrant king or giant emu or caring parent. He will probably be with us for as long as *Homo sapiens* rules the earth. Perhaps in that sense we relate to him, just as the ancient kings compared themselves with lions. T. rex is the realisation of all our monster dreams, the scariest, fiercest beast ever to walk the earth. Hail to the king!

Lazarus and the earthstar

In the Gospel of St John in the New Testament, Jesus performs his greatest miracle by raising a man from the dead. Lazarus had been dead four days and was already in his tomb, wrapped in his shroud. Jesus ordered the rock seal of his tomb removed, and out walked Lazarus, still in his shroud, all ready to resume his life.

Lazarus has lent his name to species that reappear as if by a miracle after the scientific world has declared them dead. There are a surprising number of Lazarus species out there, in the ocean depths, on islands, in remote forests and mountains, but also closer to home. Perhaps a full third of species declared extinct have since reappeared. They are one reason why we can never be absolutely sure how many species are, in fact, extinct.

Lazarus species do other things apart from returning from the dead. They give us hope, hope that a lost species may yet return. A famous example is the Bermuda petrel or cahow, named after its eerie cry. The sailors who docked in the great harbour of Bermuda back in the 1500s found the air full of long-winged birds crying their name: *cahow*. Perceiving it as a God-given source of food, like manna from heaven, they tucked into roasted petrel and boiled petrel eggs, while their hogs roamed the island's forests, rooting up the petrel nests and consuming the contents. By 1620 there were no more cahows crying over the harbour or anywhere else. It was one more island bird gone, one among many.

Then, more than three hundred years later, in 1951, they found another cahow! Admittedly it was dead, having flown into a lighthouse, but the discovery inspired a search and several nest-burrows were eventually located on uninhabited rocky islets, still within sight of the harbour. A tremendous effort was made to save the bird, involving ongoing battles against invading rats, ingeniously designed artificial burrows, and nest-caps designed to admit the birds but not their predators. It also propelled the complete restoration with native plants (themselves under threat) of a larger island, called Nonsuch, where cahows could nest in comparative safety. And the result of all this effort was that the cahow increased from 17 nesting pairs to . . . 18 (out of a total of perhaps 300 birds out in the vastness of the Atlantic, 46 birds fledged successfully between 2009 and 2016). Cahows are long-lived birds that lay only one egg per season, and do not mature until they reach three or four years old, assuming they survive that long. Restoration has to be slow. The greatest threat now is storms, which are becoming more violent by the year as a result of climate change. Storm after storm battered Bermuda in 2020, flattening the trees and stripping away the soil. As they say, it's one damn thing after another.

The cahow's story inspired a book and a film. It became a species in the spotlight, an avian parable that inspired hope and consolation. It will, I know, seem a steep descent to move from the lovable cahow to a small oddly shaped fungus growing in an English wood.

But that, alas, is where my own small experience of Lazarus species lies, in dingy, modest things that few people would even notice. My personal ray of hope resides in the humble earthstar, a fungus related to the puffball. There are fifteen species of earthstar found in Britain, not many fewer than in all Europe. But of those, nearly half were assumed to be nationally extinct on the grounds that no one had found them for a long time. Something bad, one might deduce, must have happened to earthstars.

What is an earthstar? Walking past a woodland bank on a mild autumn morning, or perhaps making your way across the back of the dunes after the crowds have gone home, you may come across a little brownish object shaped a bit like a peeled orange, or spread out like a starfish, with a little ball in the middle with a little round puncture on the top. Like a puffball, the earthstar produces puffs of 'smoke' – its dust-like spores – whenever it is hit by a raindrop. In slow, drizzling rain you can watch them puffing away like smoke signals. But unlike puffballs, earthstars can grow in dry places thanks to their tough waterproof skin. This same insulating layer enables them to carry on puffing for weeks on end until the fungus runs out of 'fuel' (i.e. its spores). You can find their empty husks in the spring, having lasted all the way through winter, through frost, snow and fierce drying winds.

If earthstars were as common as puffballs, they would undoubtedly have become part of the world of myths and fairy stories. Perhaps, if we came across a circle of little fungal stars we would routinely hop inside it and make a wish. Or use them as Christmas tree decorations, dangling on the outspread branches like exotic fruit. As it is, most of them are found only occasionally, and some are obviously rare. The unmistakable pepperpot earthstar, for example, had not been seen in Britain since 1880. The last daisy earthstar was noted just before the Second World War, and the last Berkeley's earthstar (named after the Rev. Miles Joseph Berkeley, one of Britain's numerous nature-loving Victorian vicars), a few years before that. What was depleting our earthstars? No one knew. But when, in the 1990s, Britain's first draft Red List of lost or endangered fungi was compiled, earthstars featured prominently. You looked at the list and wondered what kind of environment could fail to accommodate such tough and undemanding things as these?

As it turned out, the problem wasn't conservation so much as perception. It wasn't the earthstars that were missing so much as people who can identify them. And the reason for that was the absence of a good field guide. All it took to rescue the earthstar from the void was a book, and that appeared in 1995 with the Kew Gardens-published *British Puffballs, Earthstars and Stinkhorns* (Pegler 1995). It brought attention to the wonders of earthstar biology in the nick of time when quite a lot of naturalists were turning on to fungi.

Like Lazarus species the world over, the British earthstars were only lying low. Armed with the Kew book, zealous fungus finders started searching dry banks, churchyards, gardens, even flower beds. An earthstar fan group sprang up, operating under great secrecy, treating their quarry like rare orchids and revealing their sites to no one. And in the process, these nature detectives rediscovered not just one or two of the lost species but the whole lot, every single one of them, innocently puffing away, even the long-lost pepper-pot (it turned up on a sandy lane in Norfolk, not far from where it was last reported, a century earlier). To top that, they also found a completely new species which was named *Geastrum britannicum*, the British earthstar, because we found it first. Actually it turns out to be one of the commoner ones.

The moral seems to be: never write off a fungus. There is now a Citizen Science Lost and Found Fungi project, with its own website, dedicated to tracking down missing funguses and with an impressive track record of success. Its participants have become fungal sleuths, following clues in ancient journals and records, or even chance sightings from passers-by. With micro-fungi, in particular, an absence of records may simply indicate an absence of specialists (the few professional specialists that are left now work mainly indoors, attending meetings or filling in forms). My friend and botanical super sleuth Malcolm Storey was once employed to look for certain lost rusts and smuts, parasitic fungi that have evolved to infect just one species of plant. Among them was a rust called *Puccinia polemonii*, a species which had been unwise enough to trust its fortunes to a rare and decreasing plant, the Jacob's ladder, a tall plant with a cluster of bright blue flowers above curiously ladder-shaped leaves. On the grounds that no one had spotted it for a long time, *polemonii* had been added to the provisional Red List as critically endangered or possibly nationally extinct. So Malcolm went off to the English Peak District, found some Jacob's ladders, and peered at them very closely with his hand-lens. And almost right away he spotted the distinctive rusty pustules of *Puccinia polemonii*, not seen by the human eye for decades, and not extinct at all but merely minding its own business among the cows and buttercups of the dales. Malcolm claimed the finder's privilege of naming it and opted for 'rusty ladders'. But the committee that decides on these things doesn't like jokes, and so the fungus became known instead as the Jacob's ladder rust, and it is no longer Red Listed either.

So it is a risky thing to write off a fungus, and the same must be even truer of micro-organisms, rotifers, say, or protozoa or bacteria. Fungi are especially hard to monitor since they appear only now and again, and only for a short time, mostly in the autumn. For the rest of the year they exist only as tangled threads in the soil or embedded in wood, and so invisible to our eyes. You need to visit a place again and again to gain a good idea of what is there. All the same, the evidence does point to a real and possibly steep Europe-wide decline of fungi, including sought-after edible species such as the chanterelle or the wild field mushroom. Fungi are known to be sensitive to chemical changes in their environment, including the enormous amounts of active nitrogen produced by our cars, fertilisers and fires. Climate change adds its poisoned pennyworth and, in many places in Western Europe, edible mushrooms are becoming harder to find.

If fungi really are in trouble, we need vivid examples to draw attention to them. In that sense an extinct mushroom that also happened to be colourful and delicious would still have a useful role to play. There may be a candidate waiting in the wings (for it is not yet extinct) in *Gomphus clavatus*, the pig's-ear fungus (the equally unflattering scientific name means 'plug-like/club-shaped'). Viewed with sympathy, it is a handsome and unusual mushroom, like a piece of natural sculpture, shaped like a vase with violet-tinged flesh and the size – and shape – of a teacup (but exceptionally growing to the size of a teapot). Pig's-ears grow in clusters and even rings in natural conifer forests, usually on hillsides above two thousand feet. Since it is edible, as well as unmistakable, people look out for it. Pig's-ears have, or had, a large natural range across Europe but they are commonest in the vast natural forests of birch, spruce and fir in Scandinavia. Britain lost its pig's-ears early on; it was last seen in the 1920s. But it also seems to be dying out across Europe and is now Red Listed in seventeen countries. It is one of a small number of fungi proposed for Europe-wide protection under the Bern Convention on habitats and species. So the bells are definitely ringing for the pig's-ear. But what, exactly, are we meant to do about it?

Almost certainly this is another victim of insidious changes in the environment that has tipped the delicate balance of soil chemistry and plant physiology needed by a large fungus growing on tree roots. *Gomphus clavatus* seems to be heading for the exit, possibly quite fast. We have made a pig's ear out of it. And if it is happening to *Gomphus*, one of the most distinctive and easily recognised fungi in the world, then what may be happening to the legion of less exalted fungi that also grow on tree roots and are sensitive to change? Like the majority of living things in this world they may be doomed to disappear without anyone noticing. No funeral bells for them.

The military orchid

Let me start by asserting that 'military orchid' is a rotten name. Orchids are routinely named after the little men, women or beasties seemingly mimicked by their dangling flowers: man orchid, lady, monkey, lizard, frog, bee orchid. But 'military' is only a lazy translation of a species name, *militaris*. What we have in the flowers of *Orchis militaris* is not a 'military' but a soldier, a toy soldier uniformed in red, with a coal-scuttle helmet and two rows of little buttons running down his tunic. The orchid's early names were not only soldierly but aggressively masculine: soldier's *satyrion* – satyrs were notably priapic – or soldier's *cullions*, that is, soldier's balls, and not musket balls either.

I make the point because the British writer Jocelyn Brooke's autobiographical novelette, *The Military Orchid*, is about soldiering. Brooke had been a 'pox wallah' in the Royal Army Medical Corps, accompanying the Marines and other army units during the Second World War. Insofar as his book is about flowers, it is about nostalgic encounters with orchids in the English countryside in happier times, and, above all, the symbolic power (for him) of the rarest of them all, the military orchid. To Brooke, who grew up wanting to be a soldier, the orchid 'had taken on a kind of legendary quality, its image seemed fringed with the mysterious and exciting appurtenances of soldiering, its name was like a distant bugle call, thrilling and rather sad, a *cor au fond du bois*'. Brooke wanted to be *soldierly*.

'The idea of a soldier, I think,' wrote Brooke, 'had come to represent for me a whole complex of virtues which I knew that I lacked, yet wanted to possess: I was timid, a coward at games, terrified of the aggressively masculine, totemistic life of the boys at school; yet I secretly desired, above all things, to be like other people. These ideas had somehow become incarnated in *Orchis militaris*.'

It is necessary to add two things. The first, which Brooke could mention only in hints, because, in 1948, it was illegal, was that he was homosexual. And the other is that Brooke thought he would never find his botanical counterpart because the flower was extinct in Britain, or so it was thought. That was part of its lure. We long, do we not, for achievements we know to be impossible? Those spikes of little dangling toy soldiers seemed to have vanished, 'gone with scarlet and the pipe-clay, with Ouida's guardsmen and Housman's lancers; gone with the concept of soldiering as a chivalric and honourable calling'. Lost in itself, the orchid symbolised a lost age. It held this symbolic puissance precisely because it was no longer to be found.

In Britain, the military orchid is close to the northern and western edge of its range and was never common. It was found mainly along the edges of woods in the chalk hills around London in a narrow and ever-shifting zone between sunshine and shade. The last official sighting – 'official' because orchidophiles were always chary about publicising localities – was in 1914, when Brooke was only five. But it may well have lingered on in an unofficial secret life. He mentions 'a botanist of my acquaintance' who, 'some years ago' found a single plant of the orchid, 'carefully protected, on a private estate'. Its flowers were already fading, and, on returning a little earlier the following year, the plant had vanished, whether dug up, or rabbit-nibbled, or death from natural causes he could not say.

Where had all the orchids gone? There was a considerable ploughing of chalk downland in the later nineteenth century, especially on the lower, gentler slopes which the military orchid preferred. Its dashing appearance attracted pickers and gardeners, but the most persuasive cause of decline was an increase in rabbits, whose nibbling turned patchworks of scrub and grass into smooth, close-nibbled bowling lawns. Perhaps we could also indict the pheasant, for the woods of late Victorian England were heavily stocked, and pheasants seem very partial to orchid buds. The loss of every military orchid in England became apparent only gradually; people looked in the old places and failed to find it. And so *Orchis militaris* became Britain's first extinct orchid (but not the last; the summer lady's-tresses was last officially seen in the 1940s).

As it happened its extinction proved only temporary. By coincidence, in the very year that Brooke was writing *The Military Orchid*, it had been rediscovered. As the happy discoverer, J. E. 'Ted' Lousley, modestly recalled, 'In a way it was just luck. The excursion [in May 1947] was intended as a picnic, so I had left my usual apparatus at home and took only my notebook. But I selected our stopping places with some care, and naturally wandered off to see what I could find. To my delight I stumbled on the orchid just coming into flower . . .' (Lousley 1950). It was one of those crystalline moments in a life that obituaries love to highlight, like winning the VC: Job Edward Lousley, called 'Ted', banker and botanist, Finder of the Military Orchid.

The place was called Homefield Wood in Buckinghamshire. Clear-felling during the war had drenched the chalky ground in warm sunlight probably for the first time in decades, and in doing so had probably revived orchid seeds or tubers resting in the soil. Though he published the discovery, Lousley divulged the location to nobody. He believed that security lay in secrecy. And so it was not until the 1960s that the site emerged into the daylight when Richard Fitter and Francis Rose tracked it down using clues unwittingly left in Lousley's account. Famously, they sent him a cheeky postcard: 'The *soldiers* are at *home* in their *field*'. *Home-field* Wood, get it? The place was subsequently bought by the local wildlife trust as a nature reserve, and ten years after that it was opened to the public.

The military orchid has thrived in its home fields. From the thirteen flowering spikes counted by Lousley (plus five that had been bitten off by rabbits, plus twenty-one that hadn't bothered flowering at all) you can now find a couple of hundred. And you don't have to look very hard because every plant wears a hat, a little shelter of chicken-wire. Jocelyn Brooke's vegetable soldiers are themselves guarded by trust volunteers and their parade ground is managed as zealously as any garden. The area is divided into paddocks, winter-grazed by imported sheep; raked in places to improve the seed bed; and cut periodically to remove over-shading scrub. The orchids themselves are pollinated by volunteers using little grass blades or wires to mimic the tongues of bees. And some of the flowers are not nature's own but test-tube orchids, propagated by the Royal Botanic Gardens at Kew. Others have been dug up and transplanted to a place 25 kilometres away which, it was agreed, was in need of some military orchids too.

One way or another, the orchid's many admirers are taking pains to ensure that it won't go extinct again so easily. The species is officially out of danger in England. But in all this commendable experimentation and guardianship, has something else been lost? I have visited Homefield Wood several times over the years, and each time with diminished excitement. The plants sit inside their several cages, like exhibits. Photographers queue to snap a flagship plant from which the wire hat has been kindly doffed. An official, there to ensure nothing is trodden on or removed, explains to the visitors how the orchids have responded so well, you might almost say gratefully, to all their hard work.★

Are all the fences and gates and wire hats necessary? The nearby woods are full of deer and rabbits and pheasants. But you sense there is more to it than that. Of all wild flowers, it is the orchids, in their exotic beauty and strangeness, in their near-animal cuteness, that bring out our caring side – and the gardener within us. In England, though less so elsewhere, I have seen wild orchids cossetted inside all kinds of protective gear from little home-made wigwams to professional-looking shelters, orchids sealed off with tape as though they were the victims of an accident, orchids advertised as plant celebrities on big signs by the entrance to their reserve.

This is the fate of plants that have slipped too close to the edge. The military orchid is not endangered worldwide,

---

★ In its second English locality, discovered in 1955, the military orchid inhabits a disused Suffolk chalk pit inside a plantation. There its conservation consists of a tall fence with a locked gate and a boardwalk to stop you treading on the flowers. It is open to the public on only one day a year, and not at all in 2020 or 2021, though I am told the gate is climbable.

but what matters more is that it is vanishingly rare in England. The world is full of plants whose further survival in the wild is doubtful without intervention. Where all else fails, the solution may lie in propagation and conservatories. In effect the dying species becomes a garden plant.

I have sat next to and touched another of these lost plants in the Palm House at Kew Gardens. Its name is *Encephalartos woodii*, Wood's cycad, and it is magnificent. Its rugged trunk sprouts a fountain of elegant ladder-shaped leaves in the midst of which one or more tall orange cones project, the size of footballs. It is said to be the last representative of its kind on earth, billed as the world's loneliest tree. Cycads are either male or female. This one is male. On the Kew website, it has its own lonely-hearts ad: 'Lonely cycad seeks partner.' There are several more Wood's cycads about the world in greenhouses and botanic gardens but they are all clones of this one survivor. Unless a female partner is found, it cannot breed and so has hit the buffers as a functioning species.

Labs, conservatories and gardens are likely to be the last places of every endangered plant on earth, or at least those we choose to cultivate. Such species may well continue or even thrive under our supervision – look at the ginkgo, the 'maidenhair tree', extinct in the wild but growing happily in botanic gardens and city squares and streets all over the world. It continues to function as a species but no longer in its natural habitat. For many useful plants, trees above all, the border between the wild and the garden has been blurred. For example, who is to say which English hedgerow apple is truly wild and which the product of cultivation and planting? And how much does it matter?

I think that in the case of the military orchid it does matter. What I felt on my last visit to Homefield Wood (for I will never go there again) was that we seem to have saved it at the expense of its wildness, its birthright, so to speak. We have placed ourselves in control of the whole process; 'management' and propagation have taken over from nature. Had its conservation begun in 1860 instead of 1960 other options might have been available. We might have been able to preserve large acreages of downland and scrub and given the orchid its chance to make a showing in the wild. As it is, radical intervention was called for. And so the system cranks into gear with all its multiple actors and components – politicians, scientists, laboratories, bureaucrats, accountants, conservation volunteers – all wheeling into action in a torrent of budgets and agendas and plans. And in the process, without really meaning to, we avoid 'Extinct in the wild' and replace it with 'Extinct *from* the wild'. We have made the species our own. We have welcomed the orchid into the garden and softly closed the gate.

Here be dragons

When I was a child the school's morning assembly was a mini church service, a hymn and a prayer, prefacing the sports results, announcements and list of detentions. One of the regular hymns was almost a folk song, 'When the knight won his spurs in the stories of old', and it compared the dashing knight of yore with the modern Christian and his own battles against sin. One of its lines I found strangely sad: *'Back into storyland giants have fled / And the knights are no more and the dragons are dead.'* I wasn't bothered by the fleeing giants particularly, but there was something terribly touching about the dead dragons, their purpose gone, their existence denied. For this little child of six, some of the colour had just dropped out of the rainbow.

Dragons once existed on the same basis as other fabulous animals: griffins, for instance, or unicorns. Chroniclers mentioned them as if they were as real as lions or elephants. In winter 1016, for instance, a dragon was reported to have devastated Armenia, 'vomiting fire on Christ's faithful'. Another was seen flying overhead in Hungary by a reliable witness – a monk – who noted that its plumed head seemed as high as a mountain, with its body covered all over with 'scales like shields of iron'. Most religions of the world rest on a belief in supernatural beings. If you believed in Jupiter or Odin, why not dragons?

They also appeared in bestiaries, compendiums of real and imaginary beasts which were the medieval version of zoology. Under the heading of dragon, you would see a picture of the beast and an account of its habits, citing various learned authorities in support. Where bestiaries differed from field guides was in the conviction that every animal had a moral lesson for us. Because they were all created by God, so must they have a divine purpose, for God would not otherwise have created them. Dragons usually represented sin. They were evil, creatures of the devil. They lurked in the corners of the world, lying in wait. The lesson for the good Christian was to be watchful.

Dragons, we were told, lived in remote uninhabited wastelands. 'Here be dragons' would be written on the uncharted territories of the world, just as monsters of the deep lurked in the world's oceans and glared from blank spaces on the charts. Perhaps dragons created their own uninhabited wastes by pillaging and burning; at any rate they stood at the opposite pole from cities and civilisation.

You could, from a study of the chronicles and bestiaries, compile a fairly convincing natural history of dragons. Of course there would be problems when it came to dragon physiology. How to explain how they breathed fire without scorching themselves, and what was the fuel anyway? How could such a long, heavy and plainly non-aerodynamic body take flight, especially with such stubby, bat-like wings? But scientific inquiry wasn't necessary. Plainly dragons were magical beasts, and since magic existed, then there was no physical problem. All the same, the dragons of world myth were not always huge, and not all of them breathed flames either; some were merely venomous, like snakes. On the whole medieval dragons were more believable than their digital descendants in films like *The Hobbit* and *The Game of Thrones*. Those beasts, the size of Superfortress bombers, were probably inspired less by bestiaries than modern warfare, especially the fire-bombing of Japanese cities by B-29s in 1945. All the same, even 'real' dragons could grow very large, for, according to some accounts, their favourite prey was elephants, which they would strangle with their tails. The 'proof' that dragons existed lay in the huge and mysterious bones found buried in sediments. Of course they were not the bones of bygone dragons, but mammoths and other extinct giants had not been discovered yet while everybody knew about dragons.

A zoologist studying the natural history of dragons would probably conclude that they were not a single entity but a family of species, like crocodiles or iguanas. There were, for example, comparatively small, snake-like dragons, often called 'worms', that lived in ponds or holes in the rock. Not all dragons had wings either and, in the Far East, they were associated not with fire so much as water. Chinese dragons could cause it to rain and so were, very broadly speaking, beneficial. One of the founding myths of England is of course St George and the Dragon, in this case a malevolent winged beast that took a lady hostage and forced the town to offer up two citizens a day for its supper. The many artistic images of that fight usually show a small and rather pitiful dragon being speared by the armoured saint on his charger, no worse a challenge than sticking a pig. Perhaps they were shown this way for reasons of space or, with their shaky sense of perspective, because a big dragon was harder to draw.

A much friendlier dragon, *Y Ddraig Goch*, coloured red, symbolised the Welsh people when it fought with and overcame the wicked white Saxon dragon. It still flies on the flag of Wales. Like St George, the historical truth of the tale mattered less than its power as the founding myth of a nation. Perhaps the worst of all dragons was the one in the Book of Revelations, a terrifying beast with seven crowned heads and ten horns, an emissary of Satan. The New Testament borrowed the image from Greek and Roman mythology. Dragons have been around a long time. There seems to have been a near-worldwide need to invent such a beast, to inhabit the boundary where reality touches dreams.

Dragons were obviously culturally useful too, serving as metaphors and symbols that enlightened the world. But did people really believe in them? While history offers no clear answer, one could reasonably enquire: why not? The pre-scientific age was one of belief, not objective truth, and magic was a force that was recognised both in folk tales and, in a more circumscribed sense, in religion. Mystery could remain; it was acceptable, and Christianity itself rested on the mystery of the Trinity. Because something was incapable of rational explanation did not necessarily make it less real. In fact dragons had nearly all the attributes of a real beast except one: actuality. They remind you of the definition of a factoid, something which sounds like the truth, which has all the attributes of the truth, and is believed by many *as* the truth, but which unfortunately happens to be untrue. Many of the attributes of the truth but not the vital one.

So dragons, like other lost life, had a form of existence: in reported encounters, in beliefs and in myth. They were a presence in the world. It is in that sense that I claim the dragon as an extinct animal, and moreover one which, like the dodo and *T. rex*, possesses an active afterlife. Dragons are, in the words of the hymn, 'dead', but only because no one believes in them anymore. They offer yet another form of extinction. They remain alive in our imaginations as the films show so vividly (I bet that the amount of research devoted to the creation of Smaug or Drogon surpassed that of any dinosaur). To many of us they are among the more familiar animals on earth. That they never existed is scientifically true. But we should remind ourselves that for most of our time on earth humankind lacked the intellectual or physical means for scientific inquiry. Instead we modelled the world on our imagination, on what was believed to be true. The world was, is, and always will be, what we make of it.

Chapter 7

Rebellion

Adults keep saying: 'We owe it to the young people to give them hope.' But I don't want your hope. I don't want you to be hopeful. I want you to panic . . . I want you to act as if our house is on fire.
  Greta Thunberg, *World Economic Forum*, 25 January 2019.

In ecological terms, we are almost paradoxical: large-bodied and long-lived but grotesquely abundant. We are an Outbreak.
  David Quammen (2012), *Spillover: Animal Infections and the Next Human Pandemic.*

Nature conservation used to be so polite. I have worked in the conservation industry, in one way or another, for half my life. I wrote one of the first books on urban wild-life, and, later on, the volume, *Nature Conservation*, in the Collins New Naturalist series, as close to a British standard text as any. I know what it's like to try to halt the jugger-naut of progress to save a patch of 'scientifically interesting' ground: that is to say, quite often heartbreaking but with an occasional sense of something achieved. If it is a matter of choosing sides, I am for nature, permanently, always. Borrowing words from Dante, nature is my guide, my teacher and my lord. I am not without hope for the future, because the alternative is to despair and do nothing. All the same, nature is undoubtedly in decline the world over, both physically and figuratively, in our sense of it, in our connectedness to it.

Nature conservation, in Britain at least, operates in a kind of dream world where everyone has fine objectives, where everyone agrees that nature is important, in fact *essential*, but somehow those objectives are rarely met, and the confident targets impossible to achieve. On the positive side, we have certainly reduced our dependence on fossil fuels, and have developed an environmental conscience, both individually and as a society. But the farm birds go on declining, flying insects continue to disappear, and natural habitats such as downs, heaths and bogs go on shrinking. Profitable ways of farming and forestry have little or nothing to do with nature. Palliatives such as mass tree planting simply substitute the commonplace for the unique.

Conservation in practice was usually courteous, and on our side almost apologetic. In trying to persuade landowners to mitigate their plans in order to preserve some fragment of wild land, we came cap in hand, with little to offer beyond persuasion (while the landowner was himself being persuaded to squeeze every last morsel of profit from the land by his union, by agricultural colleges, by forestry interests, by government, and above all by his peers). In an old land like Britain where most property is privately owned by a small minority, the owner holds the cards. Partly for that reason, the framing of the law has produced a labyrinth of regulation. As red tape spooled out over desks and office floors, conservation became less to do with understanding wildlife and more about accounting, legal interpretation and internal administration (oh, those endless re-organisations) – from the outdoors to the indoors, so to speak. For reasons of their own, the government generally decides to turn everything upside down roughly once per decade. New CEOs like to do the same.

As I say, nature conservation was polite and respectable. Apart from the odd *cause célèbre*, the public were rarely much engaged. Back in the 1970s and 1980s, we thought the main problem was habitat destruction, principally through the advance of agriculture and plantation forestry but also from urban development. There was no sense then of global catastrophe caused by climate change, and renewable energy seemed more like an optional top-up. On our daily rounds, in estate offices, in conference rooms, occasionally even over the farm gate, we discussed matters quietly and reasonably, as civil servants should. In Scotland, where I worked for the Nature Conservancy Council between 1977 and 1984, there wasn't even much public support for what we were doing, in fact more of a suspicion that conservation was an unwanted English intrusion. And Conservative governments, especially, saw little need of us, a view often reflected in our budgets.

Since then, of course, there has been a sea change in public opinion, a realisation that the world is changing for the worse mainly through the burning of fossil fuels. The work of climatologists has been transmitted to the public through the medium of television, especially Sir David Attenborough's high-budget programmes, and also through the growth of the internet, smartphones and social media. I believe that eyes first began to open in 1989 when, for the first time, 15 per cent of the British people voted green. In the same year, the British way of nature conservation became federal. Nicholas Ridley, never a friendly minister, had decided that the nationwide responsibilities of the Nature Conservancy Council should be separated into new bodies in England, Scotland and Wales, each taking a different path, coloured by local priorities and aspirations. Some saw it as a weakening of influence.

Meanwhile, on the margins, there were stirrings of a new militancy. Since the seventies, Greenpeace had operated on the radical fringe although its main focus was on global challenges such as whaling and over-fishing. Then, in 1990, an American-based group, Earth First!, arrived on British soil. It combined the earth ethic of Rachel Carson and Aldo Leopold with cheerful hippy anarchism. Though there were never more than a few hundred activists, they made headlines by occupying Liverpool docks (where tropical hardwood timber was being imported) and camping out on Twyford Down, a chalk hill about to be severed in twain by a motorway.

And then, in 1995, came a major turn in the road: the Newbury Bypass. The town of Newbury lies on a busy north–south road from Southampton to the Midlands and had long been a traffic bottleneck. The plan was to build a dual carriageway of some nine miles around the western side of the town, which entailed the felling of 150 hectares of mature woodland, and also some damage to marshland along the river Lambourn, home to the tiny but soon-to-be-famous Desmoulin's snail. Under Margaret Thatcher, the government's response to increased road traffic had been simple: build more roads. The Thatcher government proposed a total of 600 bypasses and road-widening schemes, although that number was eventually whittled down to 150. At Newbury, resistance solidified under the umbrella of The Third Battle of Newbury (Battles One and Two had of course taken place 350 years earlier during the English Civil War). Up to 7,000 people marched and demonstrated and a hard core of them camped out in the woods in the bitter winter cold. Such demonstrations made a name for a wide-eyed youngster called Swampy who was arrested at a similar protest at Fairmile in Devon. Until then most eco-celebrities had hailed from the academic or television world.

I remember visiting the bypass protesters' camp on Boxing Day 1995 with gifts of whisky and shortbread. Many of the campers had gone home for Christmas. The campsite was neat and tidy. Tied to the trees were cabbalistic bunches of feathers and pebbles which gave a sense of sacred ground, perhaps even nature worship. Fastened to one tree was a love token, a teddy bear. A hard-faced lad showed me one of their tunnels, a professional job with wooden joists and panelling, and equipped with an air-pump. Other protesters would perch high up on the bare branches like giant rooks. The local birds, he told me, ignore us now. We're just part of nature. ('Yeah,' said another spokesperson, 'we love nature and all that shit.') The camps were segregated according to diet. A solitary boy sat by his tent warming his hands on a blaze of twigs. Don't go over there, I was advised. He's a *meat-eater*.

This was mass disturbance on a new scale. The Third Battle of Newbury was a very English event, a quiet battle without casualties. Of course the protesters were evicted in the end and the bypass was built. But it went way over budget, and the hiring of security guards alone cost £25 million. In consequence, environmental concerns became much more important to civil engineers and the training courses they attend. And less than two years later the incoming New Labour government cancelled what was left of the Conservative's road-building programme. A quarter century further on, new trees are growing on the scars left by the bulldozers at Newbury.

Rage

I know what it's like to march along amid a sea of faces. Mass demonstrations make you feel part of something big and meaningful. Your individuality seems to dissolve and you become one with the crowd, shedding your inhibitions and tapping into some only half-understood raw impulse. It is almost like being under a spell. I can understand why people go out of character during demos, swinging from the Cenotaph, say, or leaving graffiti on a statue. You don't necessarily need to be drunk or on amphetamines. Self-induced hormones are enough. The raw power of the crowd is enough to deaden the natural, opposite impulse for peace and quiet. You are standing up and being counted for something you believe in. For a few hours you feel part of history.

Conservation thrives on hope. Hope energises us; it is the fuel of endeavour. But the trouble with hope as a motivation is that it is fragile and frequently dashed. And so I nearly cheered when a 16-year-old schoolgirl from Sweden stood up at the World Economic Forum at Davos and told the mostly mature audience exactly what they could do with their hope: '*Adults keep saying: "We owe it to the young people to give them hope." But I don't want your hope. I don't want you to be hopeful. I want you to panic . . . I want you to act as if our house is on fire.*' Sock it to 'em,

Greta! Trust a schoolgirl to cut through the tired dogmas of the business and tell it straight and unadorned. Hope didn't prevent climate change. And it didn't prevent mass extinction either.

Greta Thunberg's apocalyptic vision of the future gave the world a different kind of energy. On 17 November 2018, 6,000 climate protesters blocked the five main bridges over the Thames in London and stopped the traffic for several hours. Seventy arrests were made. That was just the start. As we know, the following October saw a rolling series of 'uprisings' in central London over twelve days. Despite heavy rain, thousands of people took part and they brought city life virtually to a halt. Protesters glued themselves to floors, trains and buildings. Two years later a splinter group protesting on behalf of better loft insulation was blocking Britain's busiest road, the M25.

During the October 2019 demonstrations, Smithfield Market, the UK's largest wholesale meat market, was stormed and occupied. They used a fire engine to spray fake blood over the Treasury building. A mock funeral marched in procession to Downing Street and on to Buckingham Palace, where a letter to the Queen was read out. A tree surgeon dressed as Boris Johnson shinned halfway up Big Ben to unfurl a banner reading, 'No pride in a dead planet'. A professor of psychology was arrested for a 'graffiti attack'. A mural attributed to Banksy warned us, 'From this moment despair ends and tactics begin'. The police were all but overwhelmed. More than a thousand arrests were made. There were similar actions in other British cities, and also around the world, in New York and Denver, in Stockholm and The Hague, in Canberra and Brisbane, in Cork and Dublin. Most of those arrested were young. In a survey of British 18 to 24-year-olds, nearly half said they supported the protesters' actions and aims. It was, after all, about their future.

The driving force behind this impressive feat of organisation was Extinction Rebellion (XR). Launched as recently as October 2018, it represented something new in environmental politics, a 'do-it-together' movement, a bottom-up campaign of non-violent mass civil disobedience. Its founders are Roger Hallam, a Welsh radical and former organic farmer, and Gail Bradbrook, a former Green Party activist. For XR's logo they chose a stylised hour-glass within a circle, symbolising how time is running out for many species (implicitly including ourselves): 'The science is clear: we are in the sixth mass extinction event and we will face catastrophe if we do not act swiftly and robustly' (declared XR's Handbook, *This Is Not A Drill*, published in 2019). XR believes that all people have the right 'to seek redress and secure the solutions needed to avert catastrophe and protect the future'. It was, in other words, 'our sacred duty to rebel'. XR wants the government to declare a climate emergency. It wants the UK to commit to reducing carbon emissions to net zero by 2025. And it also wants a powerful citizens' assembly to make sure it happens. No excuses.

To achieve that, realistically, everyone's lives would have to be turned upside down. The route out of the climate emergency would be through 'de-growth' and, probably, a new economic system, a state-impelled alternative to capitalism. New solar and windfarms would have to be built at lightning speed. (But where would all the steel come from? And how do you make steel without using coal?) All petrol- or diesel-driven vehicles would have to be replaced by electric ones, more or less at once. We would be strongly encouraged to eat less meat. Air travel would be limited, if not rationed. More and more people would work from home. The economy would need to be placed on a war footing. It would be, in effect, a revolution. Even XR admits that it would take a miracle to achieve this.

I think it would take more than that. The paradox facing XR, or any other campaign group taking direct mass action, is that to make headlines and draw attention to your cause you risk losing the very goodwill you are seeking. Perhaps the moment when Londoners lost patience with XR was when, towards the end of the twelve-day uprisings, protesters started targeting tube trains at rush hour on London's Underground. Furious commuters hauled one man from the roof of a carriage and set on him. Another wild swipe was to target the BBC for its alleged failure 'to report on the emergency'. Some of us had been given the impression that the BBC reported little else. For the same reason XR also blockaded *The Times* and other newspapers, an action which could be, and was, construed as an attack on free speech and made hate figures of the XR protesters in Murdoch-owned papers. One could appreciate their frustration but the impression of anarchic flailing was reinforced by the performance of some of XR's spokespeople who came across as evasive, to say the least, when interviewed by the likes of Andrew Neil or Piers Morgan. A notable own-goal was Roger Hallam's claim that 'billions of people are going to die in quite short order' and that 'our children are going to die in the next ten to twenty years'. It enabled XR to be pilloried as scaremongers (but 'alarmist language works', they insisted). When, the following February, some student activists cut up the lawn at Trinity College,

Cambridge, peaceful demonstration seemed to be edging close to mindless vandalism (but it wasn't quite mindless: the College had invested heavily in oil and gas companies, and the students disapproved of that).

While every thinking person on the planet is concerned about climate change, it seems to me that the climate rebels have problems beyond the purely tactical. For one, we all of us contribute to climate change by our very existence. Every one of us burns carbon and the guilt, if such it is, is universal. Perhaps, before demanding extreme, unachievable solutions, we should honestly recognise what kind of species we are, historically and biologically. I think Elizabeth Kolbert expressed it best in her influential book, *The Sixth Extinction*: 'To argue that the current extinction event could be averted if people just cared more and were willing to make more sacrifices is not wrong, exactly; still it misses the point. It doesn't much matter whether people care or don't care. What matters is that people change the world'.

However concerned we may be as individuals, as a species, as *Homo sapiens*, we are masterful, resourceful, competitive, super predators; we are conditioned to behave like that by virtue of natural aggression, large brains and vivid, inventive minds. It was easy for climate rebels to be pilloried as hypocrites, as Piers Morgan demonstrated so brutally on one of his breakfast interviews: 'Do you have a TV? I repeat do you have a TV? Answer the question. *Do you have a TV?*' Yes, many of us own a TV, a car, a smartphone, burn up the air miles, and we shop in supermarkets where everything is cocooned in plastic. Some of us have very little contact with nature, little knowledge of it, and often little interest either. To see humanity as we are, and not as some would like us to be, seems to me a necessary first step. As David Quammen has expressed it, in ecological terms humankind is not an asset. We are an outbreak.

Another problem is the confrontational 'us-and-them' impression that rebellion creates. To do something about climate change demands a collective effort, but the sight of XR supporters dancing in the streets created, justly or not, a sense of alienation: we here are righteous and concerned, you over there, yes *you*, are the causes of the emergency. XR divides us all into 'workers' and 'the privileged' (Which am I? No idea). But, young or old, worker or privileged, we are all world-changing planet-trashers.

I admit that, like many, I feel conflicted about all this. I want the wildlife of the earth to thrive within their natural habitats and not just in a zoo or a safari park. I don't want more species to die out because of us. I would like people to be reasonable and make sacrifices to ensure a sustainable future for life on earth. But at the same time I want to retain the freedom to live as I choose. I don't trust dictatorial governments. History shows that utopian solutions always end in disaster, especially for the ruled. But I also think there is a limit to what can be achieved in a democracy. And besides, even if Britain went carbon zero tomorrow, what difference would it make to the world's climate problems? It would be a gesture, not a solution, and in any case are the Brits still so influential in the world that others look to us for an example? In the end, I suspect, humankind will just muddle along somehow, albeit in an increasingly fractious world, decreasing our emissions per capita but also increasing our numbers And who knows what will become possible, given time?

The problem for the world's wildlife is that even zero carbon economies will not help the birds, animals, insects and plants if their habitats have meanwhile been destroyed. There is an obvious link between climate change and biodiversity, for rising sea levels, fire and flood, and climate chaos, they all physically destroy life. But it is not the whole story. The science of ecology is based on the idea of *ecos*, a home for every species. Without living space, their own niche environment, a species will die. And unfortunately you can't replace natural habitats with artificial ones. Planted trees, for example, cannot replace the carbon storage capacity of natural forests, nor will they support more than a fraction of their biodiversity.

It would of course help if people had fewer children and did not live into their nineties. The demographics force economic growth. Poverty and hunger demand that natural resources are exploited to the full. We might be healthier, and use less plastic, if we grew our own vegetables and didn't subsist on junk food, but for many, convenience trumps green morality and for more people still, 'convenience' is not a lifestyle choice but instead a necessity. It might help to vote Green, or would if the British Green Party showed a bit more interest in nature. But I think we are too heavily invested in our comfortable lives to be ready for radical change. I have a mad theory that technologically advanced societies eventually wipe themselves out and that the super civilisations of science fiction and fantasy probably don't exist. It seems a shame that we will probably exterminate the world's wildlife before exterminating ourselves, but some species will survive us, and evolution can get going again on those. I'm not saying it will happen (how do I know?). Only that I doubt XR's assertion that 'this is our darkest hour'. Most of you are young. Come back in forty years and take another look.

And then, after the year of rebellion of 2019, came the plague year of Covid-19 and we found we had entirely different things to worry about.

A million species?

Although XR took its name from extinction, it does not seem to talk about it much. In its 'Declaration of Rebellion' the word isn't even mentioned; neither, for that matter, is nature. XR's preferred term is 'ecological crisis'. It refers to the worldwide loss of biodiversity, but the crisis is really our own, the possibility of large-scale loss of life through floods, fire, drought and the perils of migration. The loss of wildlife throughout the world, implicitly, is only a crisis because it is part of *our* crisis. They call it ecological, but see it through the prism of humankind's sentiments and aspirations.

All the same, XR accepts the UN's assessment of a million species under threat of extinction within decades. And let's repeat that: a *million* species, *within decades*, in other words within most of our lifetimes. That was the bleak conclusion of the landmark report of the UN's cumbersomely named Intergovernmental Science-Policy Platform on Biodiversity and Ecosystem Services, or IPBES. Chaired by a leading British atmospheric scientist, Sir Robert Watson, it found that the ecosystems on which the health of the planet depends are in a poor state. Some 145 experts from 50 different countries (with inputs from 310 more specialists) concurred that the natural spaces of the world have shrunk or been damaged over the past fifty years to the extent that a significant proportion of the world's wildlife is now in danger. 'Around one million animal and plant species are now threatened with extinction, many within decades, more than ever before in human history.' The report was careful to add, and to stress, that it is not too late. Mass extinction is not inevitable. Not if we act now. In fact we must act, because our own future is also at stake.

Extinction Rebellion interpreted all this in the starkest terms: 'Life as we know it is on the brink of collapse.' In the conservation business, we use this word *brink* a lot. Always this brink. Every year (or so it seems) some important figure will come to the lectern at a big international conference and tell us that if we don't act now, at once, it will be too late. But the words cannot be taken literally since it is never too late to act. The purpose of the brink word (the B-all and end-all of environmental conservation) is obviously to inject some urgency, some energy, into the proceedings. It pictures a great precipice over which we will topple heedlessly to our doom, like lemmings, unless we adopt new, more rigorous environmental policies that permit us to step *back* from the brink. Yet it seems that we will never actually topple over and crash because this is a magically moving precipice, forever a step or two ahead. The repeated sound-bite of 'act now before it's too late' risks a diminishing return through the loss of credibility. Unfortunately, and by a paradox, as the years go by it becomes increasingly close to the truth.

Personally, I think 'acting' now, whatever that might entail, would alleviate rather than prevent mass extinction. Who am I to disagree with the UN scientists and their million threatened species? But I do. I think we stepped over *that* brink some time ago. We are not about to witness the Sixth Extinction. We *are* witnessing it, and many species have disappeared from our lives, locally, nationally, globally. It is quite possible that a million species are extinct already. It's simple maths. All you do is scale up from the known (or partly known) extinctions of mammals, birds, reptiles and amphibians to the vastly greater numbers of plants, insects, worms, fungi and micro-organisms for which there is no data. These less-known species are declining too, and, as far as we can tell, at about the same rate. Hence from the hundreds of known extinct vertebrates of recent times you can confidently add thousands and thousands of unknown invertebrates, plants and micro-organisms. If you assume extinction happens at a constant rate across the whole of life, you end up deep into six-figures, or even touching a seven-figure total; in other words, a million species. Of course we can't name all these theoretically lost species because most of them haven't yet been found. They will have died out undetected and unknown to science. But I regard it a million or so lost species during the industrial age as a mathematical certainty.

At least as many species again are endangered, and without adequate protective measures are probably doomed (for some, like rhinos and sharks and coral, even worldwide protection hasn't done them much good). The official figures presented by the IUCN suggest as much. It believes that some 35,000 species on earth are threatened with extinction. But those are simply the species that have been sufficiently studied and monitored for their plight to be understood. Of *all* species on earth, the IUCN believes 28 per cent are in trouble. Let's scale it up again. According to the Royal Entomological Society, about one million insects have been scientifically described. But it thinks there may be another *nine* million more insects that have *not* yet been described. Assuming insects to be as threatened as any other form of life, for which there is some warrant, then 28 per cent of one million – the known insects – comes to 280,000 species. But 28 per cent of *nine* million – the unknown insects – makes 2.5 million species, and that's just the insects. Factor in spiders, mites, roundworms, flatworms, sponges, shrimps, crabs, echinoderms and other invertebrate life and we may be looking at upwards of three million species whose remaining years on this planet are looking distinctly constrained.

If I'm right, then the world has already lost a million species and will probably go on to lose several million more. Many of those on the IUCN's critical list have probably gone already, since it includes species that have not been seen for decades. Granted that lost species are sometimes rediscovered, most are probably really extinct and sleeping with the trilobites. The threat today is for the *next* million. And then the million after that.

According to the French evolutionist and philosopher Teilhard de Chardin, evolution works in a directional, goal-driven way, which he called orthogenesis. As a Jesuit priest, as well as a scientist, he believed that humankind's own evolution was foretold from the beginning, and that we are the culmination of a cosmic flow of energy directed by God. There will come a moment in the future, he thought, when the final development of the biosphere and humankind's place within it will be perfected. That is the ineffable power which, like gravity, pulls all creation towards it: the mystic union of man, nature and God. Teilhard called it the omega point, the last letter, the end time. A happy ending.

That strikes me as too hopeful. If there really is a pulling power that directs existence, it is not cosmic energy but extinction. Until humankind came along extinction was a mostly positive force. The ending of species creates new life by replacing the outmoded, the evolutionary has-beens, with new, better adapted forms based on previous models. Extinction is natural and progressive. Without it life would be stuck in a steady state and would perish entirely once the continents collide or when a random space rock hits the earth. It is extinction that makes a biological future possible. It becomes tragic only when we, as sentient beings, invest emotion in it. And, even then, we are highly selective. We may weep for a dolphin or an eagle, but who cares about a spent gnat or a bygone worm? (Though actually we should weep more for the worm for it does a more important job.)

The Sixth Extinction is of a different kind to mass extinctions past. It is indiscriminate, feckless and largely accidental in the sense that humankind is not *deliberately* exterminating species – quite the opposite. It is also eliminating species that are perfectly well adapted and have millions of years left in them. It is even *in*cidental in that, if we are honest, we are far more preoccupied with prolonging our own existence and quality of life than helping our wild neighbours (to listen to some television shows, it seems the main purpose of nature is to put a smile on our faces). All the same it is humankind that is exterminating life on earth.

Failing a complete cataclysm along the lines of the K/T impact, *we* will probably survive the Sixth Extinction (though the end of humankind might make a theatrically more convincing finale). But what kind of future lies ahead, once most of the world's wildlife has gone? Some biological writers have envisaged the ensuing age of man in terms of terrible sterility and loneliness. Edward O. Wilson believed that at the current rate of disruption one half of the higher life forms on earth will be extinct by the end of this century. He coined the word 'Eremocene' for it. Just as the Eocene meant 'dawn age', and Pleistocene the 'most recent age' of the Tertiary, so Eremocene is *eremos*, a desolate lonely place, the miserable-ist future characterised by existential and material isolation. Having exterminated all forms of advanced life on earth, humankind will rule the land with his flocks and his herds and his vast, vast cities. The scientist Felicia Smith put it another way. The way we are going, the largest animal left on earth will not only be the size of a cow. It will be a cow.

Will it happen?

I would like to say that mass extinction can be stopped and that it must be stopped and that it will be stopped. But I think one needs to be honest and to separate what one would like to happen from what one thinks will happen. Every time some institution surveys the state of the natural world, the results seem to be worse than predicted. For example, a 'major collaborative survey' led by the British Natural History Museum published in 2021 and covering 26,500 species in 70 countries, noted that, if we carry on as before, we will soon crash through the 'safe limit for biodiversity', after which everything topples (NHM 2021). Yet, it tells us, if we take action immediately on biodiversity loss, as well as climate change, then the worst may yet be averted. The Museum seems optimistic that this can happen, that indeed it will happen. My own experience in Britain, which this study calls one of the most nature-depleted countries on earth, does not fill me with the same confidence. Remedial action seems always just a step ahead.

We grasp at it, but it slips away, like the fruits of Tantalus. Effective conservation has always meant going against the grain, not only of political expediency but perhaps even of human nature.

Sir David Attenborough's 2021 *Perfect Planet* series contrasted the ecological devastation of the world – 'everything around us is collapsing' – with little heartwarming dibs of hope – some trees planted in Senegal, some animals released in Manaus, Brazil, a big solar farm in Morocco. It is all so little when it needs so much.

James Lovelock, the Gaia guy, believes that the Sixth Extinction is probably unstoppable now (see Lovelock 2006, 2009). Things have gone too far. Extinction in itself will not necessarily cause us deep unhappiness, for it is hard to be unhappy, or even angry, about what for most of us remains an abstraction. In fact, the loss of even a million species will barely impinge on human life at all. We won't miss them in the way that we miss species we have grown up with like the elm trees, the fled nightingale, or the snowshoe hare that can't find any snow. It will be more in the nature of an alarm bell warning us that something is rotten in the state of the natural world. The missing million may well have a consequential biological impact and harm the functioning of ecosystems around the world, and hence bring on further extinctions. But it will be a gradual process and take place quietly, away from the spotlight of news. Only the scientists studying it will know what it means.

What we will notice are changes in the places we know and love best: at home. Every naturalist knows what it is like to lose a familiar, well-loved butterfly or bird even if they survive somewhere else. It is like losing a friend. It also creates anxiety. The pleasure I feel at hearing the cuckoo announce the return of spring is tinged with dread at the thought that one day it will cease to return: it's not just a species but, as the poets recognised, a sound, a feeling, an hour-hand on the seasonal clock. Compared with that, a thousand or even ten thousand lost species across the world is at most the rumble of a distant storm, one you hear without getting wet.

But will there come a point when one also feels shame? I think I would feel a kind of species-shame, for my own species I mean, if poachers succeeded in wiping out the last rhino or if we witnessed the last death-throes of the Great Barrier Reef. Closer to home I would feel deep shame if the alpine flora of the Cairngorms, the nearest thing to a natural wilderness left in Britain, was lost to climate warming, as eco-modellers predict. Even closer to home, it would go hard with me if the great oaks and beeches of Savernake Forest died from drought or were thrown over by gales, and I would not feel consoled when they tell me they could always plant some new ones. What would I do about it? Nothing. Life would be just that little bit less worth living, that's all.

I have read the glad, confident reports of conservation bodies from the UN downwards, the plans for transformation, for restoration, for enhancement ('You think it's good now? Just wait!'). There will be local successes, and many of them, I hope, species saved by ingenious applications of science and technology. These will get the headlines while the remorseless grind of extinction remains in the shadows. I know some of the people that write the strategies and mission statements of the conservation industry, and I also know that they are less hopeful privately than they claim to be as professionals. In general, the more you know about nature, the less likely it is that you feel hopeful about the future. I might call it Marren's Law. For example, in a poll conducted by the American Museum of Natural History in 1998, 70 per cent of the respondents, all professional biologists, believed that we are living through a major extinction event. By contrast, it is easy to be an optimist when you know nothing. According to Dave Goulson (2021), not one MP in the current House of Commons has a degree in biology or ecology or environmental science. There are influential people, planners and local politicians among them, who claim nature and development can be reconciled by moving the nature. It looks good on the page. It is good news for developers, and very convenient all round. Just don't ask the wildlife.

Is it time for a personal Credo? I believe in the Sixth Extinction, that is, I think it is real, not that I welcome it. You could call it the great simplifier. It will strip ecosystems, lower diversity, reduce biological meaning. It may well have catastrophic consequences down the food-webs. I also believe the Sixth Extinction cannot reasonably be called a threat because it has arrived and it is happening, and it will go on happening, probably at an accelerated pace. Some of us will find reasons to deny it; none of us like it. But despite it, humankind as a whole will surely move forward on to broad sunlit uplands. It will have to because the lowlands will be underwater.

Declared extinct in 2020
The splendid poison frog
The smooth handfish
The Chinese paddlefish

The Jalpa false brook salamander
The spined dwarf mantis
The Bonin pipistrelle
The Wynberg conebush
32 species of orchids in Bangladesh and nine from Madagascar
15 fish from Lake Lanao, Philippines
Three Central American frogs with 22 more 'possibly extinct'

Released back into the wild
Guam flightless rail

Good luck on your tiny adopted island, Guam flightless rail.

Chapter 8

Islands of hope

A love of nature is a consolation against failure.
Berthe Morisot, French painter (attrib.).

Can the Sixth Extinction be stopped in its tracks? We are a long way down the extinction road already, and many scientists have concluded that climate change and extinction are both irreversible. Lost species pile up. Biodiversity targets ('Aichi targets') are not met. Environmental subsidies don't work. Some recent peer-reviewed science papers have been distinctly pessimistic. And Covid-19 has certainly further dented our confidence that the future is under our control. In fact, some would say that, from where we stand now, the future looks absolutely ghastly. But others insist that things will turn out well because *Homo sapiens* is clever, as well as wise, and in time will find solutions to anything. *Deus volent*. As God wills.

Of course there is still hope, and one's capacity for hoping depends on what we believe and read about, and where we place your trust. (Or alternatively we could rethink what we mean by hope.) In this last chapter, I focus on a few possibilities that might bring some alleviation to the Sixth Extinction. And finally on the solace that nature brings to the troubled soul whether or not the natural world is really doomed, and we along with it.

Technology to the rescue?

Saving nature by human invention seems paradoxical. Imposing naturalness by technology sounds even more so, a collision of opposites. Nonetheless, it is to developing technologies that governments are turning when seeking solutions to world warming. Climate change was, before Covid-19 hit us, the world's hottest topic and became so again in autumn 2021. That climate change and biodiversity loss are intimately linked is a given. Rising levels of greenhouse gases are trashing the life cycles of species and their interdependencies. Preventing further climate warming will not prevent further extinction because habitat destruction continues. But it would certainly help.

Great things were expected to follow from the resolutions of the UN Climate Change Conference in Glasgow in November 2021 (COP26). And, at a national level, research institutions all round the world are tackling some aspect of climate change. The Centre for Climate Repair at Cambridge, for example, is working on ideas for re-freezing the polar ice. On a more local scale, Net Zero Teesside aims to capture the carbon in industrial emissions and store it beneath the sea, although the necessary technology has not yet been developed. The next generation of power lines will not only move energy over long distances but will also store excess energy in lithium batteries, thus creating greater efficiency (but also increasing the demand for lithium, an industry which emits 15,000 kilograms of carbon dioxide for every tonne of lithium).

Some of the world's richest people are seeking solutions to the world carbon crisis. Bill Gates's way of sidestepping a climate disaster is to focus on industrial processes, such as the production of cement, plastics and steel, currently among the main sources of carbon emissions. He thinks carbon capture technologies are the way: that is, by some means of producing 'carbon-zero' steel, not least for building all those clean-energy wind-turbines. At the moment, he points out, there is no way of powering the world cheaply except by burning fossil fuels. That means some way must be found to remove the damaging gases from industrial emissions. But the shift towards a net zero carbon world economy by that route will not come soon and it won't come cheaply.

Jeff Bezos, of Amazon, has taken a different approach, offering $791 million out of a promised total 'Earth Fund' of $10 *billion* – still only a tithe of his fortune thanks to the boom in online sales during the Covid-19 lockdowns – to established environmental groups such as the US Environmental Defence Fund (EDF), the British-based WWF and the World Resources Institute, a global research body. Some of this shower of gold will go towards developing new technologies to combat climate change. EDF is planning to spend part of its share on launching a new satellite to monitor methane emissions and making the results known worldwide. The WWF is more likely to adopt 'nature-based solutions'.

Elon Musk, meanwhile, has offered to donate $100 million to whoever discovers the most promising way of removing carbon dioxide from the atmosphere. He hopes it will galvanise innovation and exercise global ingenuity. One of the more far out ideas is to use some new, giant form of air-sail craft to spray zillions of tiny reflective particles – a universe of glittering diamond-dust – into the stratosphere so that some of the sun's light will be reflected back into space. This solar 'geo-engineering' would turn the sky from blue to white.

Unfortunately it is not a one-off solution since the particles would need to be renewed regularly or the heat would increase again with renewed force. (Elizabeth Kolbert compared the white sky solution with treating a heroin addict with amphetamines. By curing one problem you substitute another.) Enthusiasts for technological solutions like these can point to the rapid development of Covid-19 vaccines as an encouraging example of what can be done with sufficient will, and assert that even agriculture might have sounded like a mad idea at first.

One can't help thinking of a line by Hilaire Belloc from 'The Microbe':

> *Oh, let us never, never doubt*
> *What nobody is sure about!**

Banking the bio

One way of avoiding extinction is to preserve the basic building blocks of a species by deep-freeze. In the case of plants these are conveniently packaged by nature as seeds. In Britain, seeds of most native plants have been collected from the wild by volunteers and stored underground at the

---

* *More Beasts for Worse Children*, from which the quote is taken, was first published in 1897.

Millennium Seed Bank at Wakehurst in Sussex. The bank also stores seeds from around the world, and, with nearly 40,000 species and between two and four million individual seeds, it is currently the largest such facility in the world. Orchids are prominent among them since many botanists concur that, while all species are equally important, the Orchidaceae is a bit more equal than the others. The seeds are 'available for research into finding sustainable solutions to global challenges'. In other words, if a plant dies out in nature, then at least seed may be available for a restoration attempt. Seed banks offer a hedge against extinction.

With animals the equivalent is tissue samples, live cells taken from endangered species that have to be stored at an even -80C° to preserve their DNA. In Britain a start has been made by Nature's SAFE (it stands for Save Animals From Extinction), billed as a cryogenic Noah's Ark. This new biobank builds on techniques used in farming and horse breeding to freeze tissue, including sperm and eggs, from endangered animals. Working with partners, including Chester Zoo, the European Association of Zoos and researchers at Oxford University, it aims to achieve for the world's rarest animals what Kew is doing for plants. At very low temperatures, it seems the samples can be preserved indefinitely. So far the lucky animals include the Asian small-clawed otter, the Colombian black howler monkey and the mountain chicken frog. 'Together', Nature's SAFE hopes, 'we can take positive action and secure a future for all life on earth'. Or make a start anyway.

Biobanks are the province of zoos and botanic gardens; the tissue samples at Nature's SAFE, for example, were taken mainly from animals that had died in British zoos. Of course, if an animal's habitat has ceased to exist then the only future for any animals biologically engineered from a frozen sample will be in a zoo, and so it will still be extinct in the wild, that is, extinct as a wild animal. Biobanks offer security against total extinction, assuming the techniques work and that the facilities remain free of accidents. But the genetic base of such small preserved samples will surely be too narrow to envisage a full-scale reintroduction into the wild. Future technology may change that. We have to wait and see.

Pleistocene Park

An emerging biotechnique designed to help a species at its last gasp is genetic engineering. The problem facing small populations is low genetic variability. These genetic 'bottlenecks' lay a species open to inbreeding and the ravages of disease, all the more deadly if the population is already stressed. Resistance to disease can be improved by translocating individuals from one population into another, smaller, one in a kind of genetic injection. By increasing variability it stiffens the survivability of a population and allows it to cope better with environmental change.

It has certainly helped the endangered Florida panther – the only surviving population in the eastern United States of a big cat more generally known as the cougar or puma. Its numbers were down to less than thirty individuals by 1995. Persecuted, squeezed by habitat loss and flattened by Florida traffic, latterly the panther's fertility was also falling. The population has been isolated from other cougars for hundreds of years and was plainly getting inbred. The panther's failing genetic health was given a boost after eight animals from the nearest population, in Texas, were captured and translocated into southern Florida. Since they are the same species, interbreeding was possible. Scientists sequencing the animal's genome have even worked out exactly how many translocations will be necessary for the maintenance of good genetic health: the addition of five fresh cats from outside every twenty years. It seems to have worked: the Florida population has risen to between 120 and 230 adult animals. The target is to establish three populations with at least 240 individuals each, at which point, it is believed, the Florida panther should be out of danger. All this would have seemed unlikely not long ago, but on present progress the target seems to be achievable. In the process the Florida panther has become one of the most intensively studied and expensively conserved animals on earth.

Using genetics to save existing species is one thing. But there is greater excitement in the thought of resurrecting extinct species. Of course, there are considerable difficulties in doing this and they increase with the genetic distance between one species and another. On current technology it would be impossible to bring back a species with no close living relatives. But it might still be possible to bring back the lost wild ox, the aurochs or at least a fairly convincing simulacrum of one, for it is genetically close to certain breeds of cattle, and indeed is probably their ancestor. The quagga of southern Africa, which resembled a zebra with reduced stripes, was hunted to extinction in the nineteenth century and is now represented only by skins and bones. But its DNA has been extracted and it showed that the quagga was very close to the existing plains zebra. Attempts are now being made to recreate at least the physical appearance of the quagga by selectively breeding back from its closest living relative, Burchell's zebra (also once thought to be extinct but now alive and well, thanks not to technology but to a revision of zebra taxonomy). It reads almost like a Kipling story: How the Zebra Lost His Stripes. The plan is to release these quagga-like animals into the ranges once occupied by the real thing, as a kind of living facsimile. Perhaps one day the technology will exist to clone real quaggas from their recovered DNA.

The project that really captures the imagination is the return of the woolly mammoth which died out thousands of years ago. Like landing a man on the moon, it represents the kind of awesome challenge that seems to energise humanity: we do it 'not because it is easy but because it is hard'. Is it even possible? On current technology probably not but there has been some progress towards that goal. Mammoth DNA has been extracted from the teeth of animals preserved in the Siberian permafrost, and part of its genome reconstructed: enough to show that the mammoth is quite closely related to the Asian elephant, more closely in fact than the latter is to the African elephant. Meanwhile Japanese scientists have managed to tease a little life into the cells of a long-dead mammoth by implanting them into the embryo cells of a mouse (suggesting another Kipling-esque drama: How the Mouse Rescued the Mammoth). A few of the mammoth cells reacted in ways that normally precede cell division although none of the cells did in fact divide. Whether this could be called bringing the mammoth's cells back to life, as the press presented it, or just a painstakingly induced burp of biochemistry, is an open question. Nothing more could be done with that cell sample and the hunt is now on for better preserved material. Does it exist? No one knows. It might.

The prodigious hurdles in their way have not prevented

scientists, especially Russian scientists, dreaming of the day when healthy herds of resurrected mammoths will once again populate the tundra of Eurasia and North America, helping to restore the grassland 'mammoth steppe'. Enough mammoths, or pseudo-mammoths, grazing those long-lost plains and fertilising it with their dung, would assist in carbon capture by enabling a deeper winter freezing of the permafrost while also insulating it from summer melting, thus preventing the release of harmful greenhouse gases. It would be 'Pleistocene Park' on an epic scale. The animals concerned would probably not be full mammoths but a sort of hybrid, Asian elephants with mammoth character-istics implanted into their genome. At best they would only *look* like mammoths (or what we think mammoths looked like). Whether they would really be fit to roam the northern tundra as fully wild animals is another imponderable. You can't help thinking that a more likely scenario would be a tourist attraction, not so much Pleistocene Park as Pleistocene Zoo with a DNA lab attached.

Of course, there are ethical difficulties with the idea of 'de-extinction'. Do we have the right to 'play God' in this way? And how wise, or even kind, would it be to bring a long-lost animal back to face the modern world, especially one as intelligent as the mammoth? There are also more practical objections. Many scientists would rather we spent limited resources on conserving living animals, such as elephants. (Imagine the irony if we gained the mammoth at the expense of the elephant!) Some would even question whether there is sufficient space left on earth for such a large, roaming (and possibly migratory) animal; certainly not on the Great Plains of North America and maybe not even in Siberia either. Mammoths would have depended on particular gut microbes to digest their food; are they still around or did they perish with their host? And even assuming the project succeeded, would not a herd of ivory-bearing mammoths become a magnet for the world's poachers? Some claim that the resuscitation of the mammoth would be a form of amends: a kind of environmental apology. We, who probably killed off the mammoth, have used our technology to bring it back to life: so let's call it quits. But that too would be problematic. For if extinction is not final after all, does that make it more acceptable?

The wastelands

Even as the naturally wild places of the world fragment and shrink, alternative habitats present themselves in places which industry has used up and moved on: quarries, tips, landfill, or even just bare ground where the factory used to be. These are the 'islands of abandonment', places perceptively evoked by the writer Cal Flyn (2021) where, for one reason or another, people no longer live. Such 'islands' are everywhere in the developed world, even within cities. Crowded, post-industrial Britain has a great many, for example along the Thames estuary where natural habitats have almost ceased to exist, and some of them are now nature reserves. Just as air rushes into a vacuum, so nature can be quick to re-establish vacant ground. The wastelands become bushy and coloured with wild flowers, generally a mixture of natives and hardy garden escapes. They begin to hum with bees and chirp with grasshoppers, and an ecosystem builds up that is a kind of echo of the wild. For example, on the former airfield on England's Greenham Common, where cruise missiles once stood ready in their concrete bunkers, there is now a dense intermingling of acid heath and limestone grassland, habitats that would normally be miles apart. The limestone comes from the crumbled remains of concrete, the heath from the gravel underneath. Quite possibly the Common supports more plants and insects now than the original one ever did. The downside is that however species-rich such places can become, they are often ripe for redevelopment. The waste-land in a loop of the Thames at Swanscombe is earmarked to become a major recreation resort despite it supporting as many rare invertebrates as any place of comparable size in the whole of England.

On a much larger scale we have sites of human-made disasters which are no longer fit for human occupation. Radioactive leaks, for instance, can offer great opportunities for wildlife. Those who built the nuclear reactor at Chernobyl, in what is now the Ukraine, claimed that the chance of serious core damage was once in ten thousand years. In fact, it turned out to be once in about twenty years, and once was enough. When Chernobyl blew up, in April 1986, and the nearby city of Pripyat evacuated, an area about the size of Luxembourg was left to nature. Though called the Dead Zone, or the Zone of Alienation, it now teems with life. Three-quarters of the area is now natural forest of birch, maple and poplar, sprung up over former farmland. Wolves, bears, elk, wild horses, wild boar, beaver roam and European bison was introduced into the area. Since their meat will not be fit for human consumption for a couple of centuries at least, the animals are not hunted much. Although wildlife, as well as people, perished in the disaster, and the mutation rate leapt up, nature came creeping in to heal the void. The Chernobyl Dead Zone now has one of the best functioning ecosystems in Europe, with megafauna, top predators and a biodiversity that increases by the year.

There are other such places around the world, on a smaller scale but still large enough to act as wildlife refuges. In the 140-square-mile exclusion zone around Japan's Fukushima reactor, torn apart by a tsunami in 2011, camera traps have revealed the presence of raccoon dogs, wild boar, serow (a wild goat-like animal), snow monkeys and Japanese badger, all looking healthy enough. Where people returned nervously, in 2019, to the ruins of a nearby town, the animals stared at them as though wondering what kind of creature they were. In North America another contaminated wasteland close to Denver, called the Rocky Flats, is now officially a nature sanctuary. Once the location for an arsenal of chemical weapons and a storage facility for plutonium, the place was closed down in 1989 after the leakage of radioactive waste. Although measuring only eight square miles, it is big enough to hold a 300-strong herd of elk, a wonderful diversity of land and water birds and 630 species of plants.

James Lovelock (2007) suggested that we might actually *choose* to contaminate land to keep people out and make it safe for wild animals. No one yet seems to have taken up his wise counsel, but there will no doubt be other accidents. Personally I have hopes of the Lake District where the managers of Sellafield nuclear reactor wish to bury radioactive waste in bunkers below the National Park, far beneath the pattering feet of its nine million annual visitors. Although only half the size of Chernobyl, the Lake District would make a very satisfactory wildlife refuge once it had recovered from being over-run by tourists and sheep. Big enough for a few wolf packs, for instance, and, who knows, perhaps even some mildly radioactive bison and bears too.

One hesitates even to mention the vast wastelands that would open up after a nuclear war. This is supposed to be a *hopeful* chapter.

Failing fertility

Is the human race failing? In the context of mass extinction, a diminishment in the numbers of our own species would be a hopeful sign, maybe *the* most hopeful sign. In the Western world, at least, the population has indeed tipped into a discernible fall. We are told that male fertility started to fall in the 1970s and continues to fall. By now the average sperm count has plummeted by more than half (59 per cent). At that rate, concludes epidemiologist Shanna Swan (2020), *Homo sapiens* could eventually become an endangered species. Without immigration the population of most countries in Western Europe would probably continue to decline. The causes of falling fertility are disputed. Some think it is due to an unhealthy diet, others blame toxic chemicals entering our bodies via plastics, cosmetics and pesticides. Perhaps we also damage our health by eating, drinking and smoking too much. Or could it be an evolutionary adaptation to smaller families conceived later in life? There are other signs of reproductive failure: contemporary boys have more genital abnormalities and girls suffer more miscarriages. Everything about our falling fertility is of course vigorously disputed. All we really know is that it is falling.

Most of the data indicating biological failure is confined to the Western world, that is, Europe, North America and

Australasia. For other parts of the world, where the population is still rising, and often rising fast, there is insufficient data to indicate any possible demographic slowdown. But perhaps, at least for the optimists, there is a faint hope that the world will not become quite as crowded and unsustainable as many fear.

Challenging words

The World Economic Forum's 2020 report on Global Risks ranks biodiversity loss and ecosystem collapse as one of the top five threats facing humanity. Ecosystem collapse has been compared to a flat soufflé or a burst football. The basic form remains but structure and function are weakened: species are shed, and the whole thing deflates. Although the Forum seems to regard this as a 'risk', that is, as something that might happen rather than one that is already happening, it still believes that ecosystem collapse is imminent, within decades. Apart from losing wildlife, such a scenario will also lose money. The lost tourism alone they account in billions. For the World Economic Forum, wealth is a powerful reason to preserve wildlife.

Its words are repeated in one fashion or another through national forums and local forums, all the way down to my local Area of Outstanding Natural Beauty (AONB), where we once earnestly discussed the threat to polar bears even though we do not actually have any. Everyone agrees about the solution: a global low-carbon economy which is also resource-efficient and where everybody works to a common agreement and cooperates. While this vast, over-arching aim is so far largely confined to words, we can take some comfort from the fact that they are at least the right words.

COP26

Climate change can only be kept within controllable limits if all the world agrees to participate. Since 1995, the United Nations has hosted an annual conference on climate change known as COP (Conference of the Parties). COP26, held in Glasgow in November 2021, was (of course) the 26th such meeting. It was supposed to have been held the previous year but was postponed because of the Covid pandemic.

Only three of the previous 25 COP meetings have been of lasting significance. At Kyoto, Japan, in 1997, the parties accepted the need to reduce their emissions of greenhouse gases, principally carbon dioxide and methane. Expectations were high for COP15 at Copenhagen, Denmark, in 2009, but the conference collapsed in acrimony. At COP21, held in Paris in 2015, the parties did at least agree to try and keep the rise in global temperature to within 2 degrees Centigrade. By then the consequences of a failure to do so were becoming clear: a rise in sea level, drowning various capital cities and island territories, a global epidemic of wildfires, land rendered uninhabitable, and an acceleration of mass extinction.

COP26 may well prove to be another milestone. Boris Johnson, the British Prime Minister, certainly showed the zeal of a recent convert by telling the world that 'it's one minute to midnight and we need to act now!' Many developed nations pledged to cut emissions to 'net zero' by 2050. Over 100 countries agreed to cut emissions of methane by 30% by 2030. 120 countries pledged to end deforestation. Conference agreed on a $100 billion fund to help developing countries move from fossil fuels to green energy. In most cases these pledges and agreements were not legally binding. India's surprise promise to cut emissions to net zero by 2070, for example, passed the problem on to the next generation, although its further promise for the country to get half of its energy from renewables by 2030 did sound like a serious commitment.

Conference President Alok Sharma announced that 'the end of coal is in sight', which seemed wildly optimistic given that neither China, India, USA nor Australia intends to end their reliance on coal just yet. But 40 countries did agree to phase out coal consumption, and also to push for electric engines in heavy goods vehicles (if such proves possible). If every country remains true to pledges given at Glasgow, it is estimated that the world's temperature will rise by only 1.8 degrees (though these figures were immediately disputed), rather than the 2.7 predicted on present trends. A 200-page text summarised the commitments of the parties and sets out how carbon markets are policed so that no nation is disadvantaged.

Protesters outside in the Glasgow rain dismissed the conference as 'greenwashing'. Presidents Xi of China, Putin of Russia and Bolsonaro of Brazil (representing a large proportion of the earth's land mass) failed to turn up. The future will show whether or not COP26 was the point at which the present generation commits itself to make life-changing sacrifices for a better future. What it undoubtedly did do was to grab the world's attention, and forced sometimes reluctant politicians to commit themselves to climate policies. In that sense, we can justifiably find reasons for hope in COP26.

Solace by the river

If none of these 'islands of hope' totally convince you, why not try this piece of advice from Sir David Attenborough? 'Sit down for ten minutes out of doors', he told the Call of the Wild podcast. 'Don't move. Keep quiet. Wait ten minutes. You'll be very surprised if something pretty interesting doesn't happen.' And you might also find yourself feeling a bit less stressed, a bit happier, a bit more interested. It's better than aspirin. What else will work in ten minutes? It's the Nature Cure: nature's way.

During the first Covid-19 lockdown, in spring 2020, two friends and I wrote daily notes based on exercise walks in our respective neighbourhoods. They were published later that year as *The Consolation of Nature: Spring in the Time of Coronavirus*. It was surprisingly therapeutic, a way of using the lockdown to intensify life, not depress it. We witnessed day by day the unfolding drama of the spring, one of the warmest and sunniest in living memory. One felt it all the more keenly because of our enforced confinement. I found wonderful corners near my home I had never seen before, things I'd never noticed: the hulk of a great ash tree on the border of the parish; a beautiful flowery bowl of a valley not even hinted at on the map; a quiet stretch of the river brimming with snow-white flowers. We did our best to turn each day's walk into a story that we could share with others. But, of course, it was nature that performed and we that merely reported. It became all-absorbing, those brief encounters with foxes, wild deer, songbirds, butterflies, glimpses into other lives untroubled by Covid-19. It is something anyone could do, anywhere and at any time, and, we wondered, why don't we do this more often? Why don't we simply look at things and enjoy them, just as a fisherman enjoys his peaceful time by the water, irrespective of what he catches, or a boy who stares at the stars with binoculars for the first time and thinks, why haven't I done this before? Why don't we let a little nature into our lives?

Looking and learning, one notices change too, and not just seasonal change. If I was able to compare the present wildlife of my parish with that of a century ago, I would expect to find a significant drop in biodiversity. For example, the local woods back then were coppiced and open and full of butterflies. There are none now except along the tracks. The landscapes described by the English nature writer Richard Jefferies in *Wood Magic* and *Bevis* lie close by, but Jefferies would have a hard time recognising them now. But a visitor coming to it fresh might consider them still unspoilt. We should enjoy things as they are, not what they were, or what they might come to be.

Besides, change can be stimulating. We might not like climate change, it does not suit us, but in general *Homo sapiens* likes change. Change is dynamic, forging onwards. Change is what we do to the world. In my own tiny corner of the universe, I watch species disappear and others appear. Today, I see red kites, dazzling little egrets and porcelain-like mandarin ducks, all beautiful and inspiring, and all recent arrivals. Tomorrow, I confidently expect to see cattle egrets in the fields, assuming we still have any cattle. In tomorrow's tomorrow, I would not be surprised to find that a family of beavers had made their home here, down among the floodlands of the river.

We don't need to feel we are carrying the burden of the world on our shoulders (though God knows, I've met conservationists that do). Climate change and the Sixth Extinction are too big to contemplate except in almost abstract terms. Our sympathies are for individual species we can relate to, such as the persecuted rhino, or the doomed island birds, or the poor dolphin of the River Yangtze, not the mathematical perplexities of biodiversity loss. The dying of the living world is too big to comprehend; monstrous, unforgivable but ultimately unfathomable. But *dum spiro spero*, 'while I breathe I hope'. As Voltaire suggested at the end of *Candide*, we can't do very much, individually, to change the world, but we *can* look after our gardens, our very own patch of the planet.

Care for nature begins at home.

Look after it, love it, let it live, and much, much else may follow.

References

Agusti, Jordi & Anton, Mauricio (2002) *Mammoths, Sabertooths and Hominids. 65 million Years of Mammalian Evolution in Europe.* Columbia University Press, New York.

Avery, Mark (2014) *A Message from Martha. The Extinction of the Passenger Pigeon and its Relevance Today.* Bloomsbury, London.

Barkham, Patrick (2020) 'Return of the Burbot: "Great Lost Fish" To Be Reintroduced to the UK.' The *Guardian*, 6 March 2020.

Beebee, Trevor (2018) *Climate Change and British Wildlife.* Bloomsbury Wildlife, London.

Benton, Michael J. (2015) *When Life Nearly Died. The Greatest Mass Extinction of All Time.* Thames & Hudson, London.

Black, Riley (2018) 'How We Elected *T. rex* To Be Our Tyrant Lizard King.' *Smithsonian Magazine*, 21 June 2018.

Blencowe, Michael (2021) *Gone. A Search for What Remains of the World's Extinct Creatures.* Ivy Press, London.

Boulter, Michael (2002) *Extinction. Evolution and the End of Man.* Fourth Estate, London.

Brooke, Jocelyn (1948) *The Military Orchid.* Bodley Head, London.

Cameron, Robert (2016) *Slugs and Snails.* New Naturalist No. 133, Collins, London.

Cardoso, Pedro et al. (2020) 'Scientists' Warning to Humanity on Insect Extinctions'. *Biological Conservation*, 242, 1-11.

Carrington, Damian (2019) 'Plummeting Insect Numbers "Threaten Collapse of Nature".' The *Guardian*, 10 February 2019.

Carson, Rachel (1962) *Silent Spring*. Houghton Mifflin, New York.

Ceballos, Gerardo, Paul R. Ehrlich & Peter Raven (2020) Vertebrates on the Brink as Indicators of Biological Annihilation and the Sixth Extinction. *PNAS* (Proceedings of the National Academy of Sciences of the United States), 117 (24), 13596–602.

Crates, Ross et al. (2021) 'Loss of Vocal Culture and Fitness Costs in a Critically Endangered Songbird.' *Proceedings of the Royal Society B*, 288 (1947).

Cuppy, Will (1941) *How to Become Extinct*. Reprinted by Nonpareil Books, New Hampshire, in 2008.

Darimont, Chris T., Caroline H. Fox, Heather M. Bryan & Thomas E. Reimchem (2015) The Unique Ecology of Human Predators. *Science*, 349, 858–60.

Dawson, Ashley (2016) *Extinction. A Radical History*. OR Books, New York and London.

Diamond, Jared (1986) 'Biology of Birds of Paradise and Bowerbirds.' *Annual Review of Ecology and Systematics*, 17, 17–37.

Drake, Nadia (2015), interview with Elizabeth Kolbert. 'Will Humans Survive the Great Sixth Extinction?' *National Geographic*, 23 June 2015.

Eldredge, Niles & S. J. Gould (1972) 'Punctuated equilibrium: An alternative to phyletic gradualism'. In: T. J. M. Freeman (ed), *Models in Palaeobiology*. Schopf, San Francisco.

Fagan, Brian (2004) *The Long Summer. How Climate Changed Civilization*. Granta Books, London.

Farrell, Clare et al. (eds) (2019) *This Is Not a Drill. An Extinction Rebellion Handbook* (2019). Penguin Random House, London.

Fisher, James (1967) *Thorburn's Birds*. Ebury Press/Michael Joseph, London.

Flannery, Tim (2001) *The Eternal Frontier. An Ecological History of North America and Its Peoples*. William Heinemann, London.

Flyn, Cal (2021) *Islands of Abandonment. Life in the Post-Human Landscape*. William Collins, London.

Friedlingstein, P. et al. (2020) 'Global Carbon Budget 2020, Earth System Science Data.' *Earth System Science Data*, 12, 3269-3340.

Fuller, Errol (1999) *The Great Auk*. Harry N. Abrams Inc., New York.

Fuller, Errol (2000) *Extinct Birds*. Oxford University Press.

Fuller, Errol (2002) *Dodo. From Extinction to Icon*. HarperCollins, London.

Gates, Bill (2021) *How to Avoid a Climate Disaster. The Solutions We Have and the Breakthroughs We Need*. Allen Lane, London.

Gould, Stephen Jay (1989) *Wonderful Life. The Burgess Shale and the Nature of History*. Norton, New York.

Goulson, David (2017) 'Warning of "Ecological Armageddon" after Dramatic Plunge in Insect Numbers.' The *Guardian*, 18 October 2017.

Hawksworth, David L. (ed.) (2001) *The Changing Wildlife of Great Britain and Ireland*. Taylor & Francis, London and New York.

He, F. et al. (2020) Combined effects of life-history traits and human impacts on extinction risk of freshwater megafauna. *Conservation Biology*, 34, 1-10.

International Union for the Conservation of Nature (2021) 'Red List of Threatened Species.' www.iucnredlist.org

Kolbert, Elizabeth (2014) *The Sixth Extinction. An Unnatural History*. Bloomsbury, London.

Kolbert, Elizabeth (2021) *Under a White Sky. The Nature of the Future*. Bodley Head, London.

Leakey, Richard & Roger Lewin (1996) *The Sixth Extinction. Biodiversity and Its Survival.* Weidenfeld & Nicolson, London.

Lousley, J. E. (1950) *Wild Flowers of Chalk and Limestone.* New Naturalist No. 16, Collins, London.

Lovelock, James (2006) *The Revenge of Gaia. Why the Earth is Fighting Back – and How We Can Still Save Humanity.* Allen Lane, London.

Lovelock, James (2009) *The Vanishing Face of Gaia. A Final Warning. Enjoy It While You Can.* Allen Lane, London.

MacArthur, Robert & E. O. Wilson (1967) *The Theory of Island Biogeography.* Princeton University Press.

Maclean, Norman, ed. (2010) *Silent Summer. The State of Wildlife in Britain and Ireland.* Cambridge University Press.

MacPhee, Ross (1999) 'Remember the Islands.' *Swift as a Shadow. Extinct and Endangered Animals.* Houghton Mifflin, New York.

McCarthy, Michael (2016) 'Apocalypse Unseen.' *British Wildlife*, 28(1), 30–31.

McCarthy, Michael, Jeremy Mynott & Peter Marren (2020) *The Consolation of Nature. Spring in the Time of Coronavirus.* Hodder Headline, London.

McKibben, Bill (2019) *Falter. Has the Human Game Begun to Play Itself Out?* Henry Holt and Company, New York.

Marren, Peter (1999) *Britain's Rare Flowers.* Poyser Natural History, London.

Martin, Paul S. & Richard G. Klein (eds) (1984) *Quaternary Extinctions. A Prehistoric Revolution.* University of Arizona Press.

Martin, Paul S. (2005) *Twilight of the Mammoths. Ice Age Extinctions and the Rewilding of America.* University of California Press.

McKie, Robin (2002) 'Trawling Puts Deep-Sea Fish in Danger of Extinction.' The *Guardian*, 17 February 2002.

Monbiot, George (2013) *Feral. Searching for Enchantment on the Frontiers of Rewilding.* Allen Lane, London.

Monbiot, George (2020) 'Population Panic Lets Rich People off the Hook for the Climate Crisis They Are Fuelling.' *The Guardian*, 26 August 2020.

Nelles, David & Christian Serrer (2021) *Small Gases, Big Effect*. Allen Lane, London.

Pegler, D. N., T. Laessoe & B. M. Spooner (1995) *British Puffballs, Earthstars and Stinkhorns. An Account of the British Gasteroid Fungi*. Royal Botanic Gardens, Kew, Surrey.

Peterson, Roger Tory (1970) *North American Wildlife in Danger*. Brooke Bond Picture Cards, Canada.

Platt, John R. (2021) 'What We've Lost: the Species Declared Extinct in 2020.' *Scientific American*, 13 January 2021.

Quammen, David (1995) *The Song of the Dodo. Island Biogeography in an Age of Extinctions*. Scribner, New York.

Quammen, David (2012) *Spillover. Animal Infections and the Next Human Pandemic*. W. W. Norton and Company, New York.

Randle, Zoe and others (2019) *Atlas of Britain and Ireland's Larger Moths*. Pisces Publications, Newbury.

Raup, David M. (1992) *Extinction. Bad Genes or Bad Luck?* Oxford University Press.

Sagan, Carl (2006) *The Varieties of Scientific Experience. A Personal View of the Search for God*. Penguin Press HC, London.

Sanchez-Bayo, Francisco & Kris A. G. Wyckhuys (2019) 'Worldwide Decline of the Entomofauna. A Review of its Drivers.' *Biological Conservation*, 232, 8–27.

Scott, Michael (2020) 'Alpine Blue-Sowthistle and the Challenge of Upland Management.' *British Wildlife*, 32(1), 12–18.

Scott, Peter (1963) *Wildlife in Danger*. Brooke Bond Picture cards.

Shaw, Philip & Des Thompson (2006) *The Nature of the Cairngorms. Diversity in a Changing Environment*. Scottish Natural Heritage, Edinburgh.

Smith, Felicia (2018) 'Body Size Downgrading of Mammals over the Late Quaternary.' *Science* (300, 6386), 310–13.

Stroh, Peter, and others (2019) *Grassland Plants of the British and Irish Lowlands. Ecology, Threats and Management.* Botanical Society of Britain and Ireland, Durham.

Swan, Shanna H. & Stacey Colino (2020) *Count Down. How Our Modern World Is Threatening Sperm Counts, Altering Male and Female Reproductive Development and Imperiling the Future of the Human Race.* Scribner, New York.

Thomas, Jeremy A. (2016) 'Butterfly Communities Under Threat.' *Science*, 353, issue 62296, 216–18.

Wake, David B. & Vance T. Vredenburg (2008) 'Are We in the Midst of the Sixth Mass Extinction? A View from the World of Amphibians.' *Proceedings of the National Academy of Science*, 105 (Supplement 1), 11466–73.

Walker, K. J., P. A. Stroh & R. W. Ellis (2017) *Threatened Plants in Britain and Ireland.* Botanical Society of Britain and Ireland, London.

Wallace-Wells, David (2019) *The Uninhabitable Earth. Life After Warming.* Penguin Random House, London.

Westwood, Brett & Stephen Moss (2015) 'Burbots. Unfamiliar yet Fascinating.' *Natural Histories. 25 Extraordinary Species That Have Changed Our World*, 103–14. John Murray for BBC Radio 4, London.

Weymouth, Adam (2014) 'In Search of the Last Wolf of Scotland.' *British Wildlife* 26(1), 20–22.

Wignall, Paul B. (2019) *Extinction. A Very Short Introduction.* Oxford University Press.

Wilson, Edward O. (1987) 'The Little Things That Run the World. (The Importance and Conservation of Invertebrates).' *Conservation Biology*, 1(4), 344–6.

Wilson, Edward O. (2003) *The Future of Life.* Vintage Books, London.

Wilson, Edward O. (2016) *Half-Earth. Our Planet's Fight for Life.* Liveright Publishing Corp., New York.

Yalden, Derek (1999) *The History of British Mammals.* Poyser Natural History, London.

Online references

'Ammonites Feeding.' See Blogs.reading.ac.uk/exploiters-and-exploited/2011/11/15/ammonites-feeding/

Andrews, Peter (2020) Notes and Views of the Large Copper and the Lost Fenlands. www.Dispar.org/reference

'Animal and Plant Species Declared Extinct between 2010 and 2019, The Full List.' See Lifegate.com/extinct-species-list-decade-2010-2019

'Can tech help fight climate change? Five innovations making a difference today.' See Kaspersky.co.uk/blog/secure-futures-magazine/20110

'Caribbean Monk Seal.' See Wikipedia.org/wiki/Caribbean_monk_seal

'Climate Change Is Accelerating The Sixth Extinction.' See Iberdrola.com/environment/climate-change-endangered-species

'Climate Change: Seven Technology Solutions That Could Help Solve Crisis.' 11 September 2020. See News.sky.com/story/climate.change

'Do Bacteria Ever Go Extinct?' See Phys.org/news/2018-07-bacteria-extinct-bigtime

Evans, Monica (2019) 'High-tech Hopes for New Zealand's Big Conservation Dream.' See News.mongabay.com/2019/06/

'Extinction Rebellion.' See Wikipedia.org/wiki/Extinction_Rebellion

'Last of the Giants: What Killed off Madagascar's Megafauna a Thousand Years Ago?' *The Conversation.* 29 March 2019. See Theconversation.com/last-of-the-giants-112672

Lindsey, Rebeca & LuAnn Dahlman (2021) *Climate Change: Global Temperature*. See Climate.gov/news-features/understanding.climate/

'List of Extinct Bird Species Since 1500.' See Wikipedia.org/wiki/list_of_extinct_bird_species_since_1500

This and the following lists are based on IUCN data, 2016.

'List of Recently Extinct Insects.' See Wikipedia.org/wiki/list_of_recently_extinct_insects

'List of Recently Extinct Invertebrates.' See Wikipedia.org/wiki/list_of_recently_extinct_invertebrates

'List of Recently Extinct Plants.' See Wikipedia.org/wiki/list_of_recently_extinct_plants

Louca, Stilianos et al. (2019) 'A Census-based Estimate of Earth's Bacterial and Archaeal Diversity.' PLOS Biology, 17(2) Doi.org/10.1371/journal.pbio.3000106

'Madagascar Minister Calls Protection Areas a "Failure", Seeks People-Centric Approach.' 20 August 2020. See News.mongabay.com/2020/08

'Male Infertility Crisis.' See Wikipedia.org/wiki/

Natural England (2010) *Lost Life: England's Lost and Threatened Species*. www.naturalengland.org.uk

Natural History Museum (2021) Tracking and predicting biodiversity damage. Nhm.ac.uk/discover/tracking-and-predicting-biodiversity-damage.

Newman, Rob (2019) *The Extinction Tapes*. BBC Sounds. See Bbc.co.uk/sounds/series/m0009yxd. His account of the loss of the baiji is extraordinarily powerful and moving.

NOAA. *National Centers for Environmental Information. Global Climate Report – Annual 2019*. See Ncdc.noaa.gov/sotc/global/201913

Phelen, Ryan. *Revive and Restore*. See reviverestore.org

Population Matters. *Welcome to the Anthropocene*. See Populationmatters.org/campaigns/Anthropocene

'Recognising the Quiet Extinction of Invertebrates.' *Nature Communications*, 3 January 2019. See Doi.org/10.1038/341467-018-07916-1

Roser, Max, Ritchie, Hannah & Ortiz-Ospina, Esteban (2013, updated 2019). World Population Growth. Ourworldindata. org

Scottish Natural Heritage (2015) *Climate Change Impacts on the Vegetation of Ben Lawers.* See Researchgate.net/publication/280092910_climate_change_impacts

'Scottish Snow Patches Report 2019/2020.' See Rmets.org/metmatters/Scottish-snow-patches-report-201922020

'Should We Try To Halt Extinction?' See Bbc.co.uk/news/science-environment-30916690

'The Cosmic Lottery. As Stephen Gould Sees It, Human History Is a Tale of Random Luck.' *Los Angeles Times*, 28 November 1989. See Latimes.com/archives/la-xpm-1989-11-28-vw-322-story

'Tyrannosaurus in Popular Culture.' See Wikipedia.org/wiki/tyrannosaurus-in-popular-culture

'What We've Lost: The Species Declared Extinct in 2020.' *Scientific American*, 13 January 2021. See Scientificamerican.com/article/

'White Rhino Sudan Dies: Is All Hope Lost for this Subspecies?' See Nhm.ac.uk/discover/news/2018/march/the-future-of-white-rhino

'Wolly Mammoth Comeback? 5 Ethical Challenges.' See Livescience.com/40263-ethics-of-bringing-mammoth-back

World Economic Forum (2020) 'One Fifth of the World's Ecosystems Is in Danger of Collapse – Here's What It Might Look Like.' See Weforum.org/agenda/2020/11/ecosystems

Yong, Ed (2017) 'New Zealand's War on Rats Could Change the World.' See The atlantic.com/science/archive/2017/11

Acknowledgements

After They're Gone was first conceived in Mon Plaisir, the venerable French restaurant in London's Covent Garden, where I shared a convivial table with three old friends, Tim Birkhead, Jeremy Mynott and Michael McCarthy. They told me I should write a book about extinction, gave me the confidence to attempt it, and were a selfless source of encouragement and inspiration during the research and the writing. We will, I hope, share the same table after the book is published.

Most of the book was written during the lockdown in winter 2020/21 when one's laptop was the main link with the world. In other circumstances I might have based the book on interviews and written it journalistically. As it is, I had to base it on my own accumulated experiences, and on what I could glean from my small library and what I could distil from the internet. Fortunately, the internet sources for prehistory and extinction are now as rich as any library. In terms of travel in search of rare and diminishing wildlife, I must acknowledge the knowledge, wisdom and good companionship of many friends, but chiefest of all Bob Gibbons, the best naturalist I have ever met, and whose friendly interest led me to discuss the book's outline with him, and whose comments I greatly value. I also thank

Jane Smart, IUCN's Director of Global Species Programme, for keeping me abreast of the Red-listing of diminishing plants and animals.

Various friends have read and commented on chapters, notably Tim Birkhead (on the Great Auk and birds generally), Sarah Fowler (on sharks), Mike Scott (on mountain flora), Richard Fortey (on palaeontology), Paul Raven and Jeremy Mynott (on prehistory), Emma Garnett (on environment issues) and, more generally, Steve Berry, David Pearman and Richard Price. I thank all of them. Any remaining mistakes are entirely my own.

Over at Hodder, I thank Izzy Everington for her eternal positivity, kindness and encouragement. I also thank Kwaku Osei-Afrifa for their thoughtful comments and editing, Purvi Gadia for her meticulous work in proof management, and Alice Morley (marketing), Eleni Lawrence (publicity) and Natalie Chen (design). I also thank Aruna Vasudevan for her discriminatory and thorough copy-editing, Richard Rosenfeld for his detailed and methodical indexing and Ross Jamieson for his expedient and excellent proofreading.

I could only dedicate this book to the three friends who gave me the confidence to write on such a daunting and fateful subject: Tim, Jeremy and Michael.

Index